FLIGHT SCHOOL MARKETING HANDBOOK

FOR OWNERS & PROFESSIONALS

SHANE D JORDAN

1ST EDITION

PILOT TRAINING MARKETING

Fl ight School Marketing Handbook

Turtle Creek Enterprises, LLC
d/b/a Pilot Training Marketing

WHO IS THIS FOR?

Welcome to the definitive guide for mastering digital marketing in the specialized world of aviation training. We designed this book for flight and ground school operators, aviation industry marketing professionals, and entrepreneurs who are eager to take their businesses to new heights. Whether you're at the helm of a well-established aviation training company or just laying the groundwork for your venture, this resource aims to equip you with the strategies, best practices, and actionable insights needed to soar above the competition.

The landscape of aviation training is complex, subject to stringent regulations and often unacknowledged by mainstream marketing theories. Traditional educational institutions have long embraced digital marketing, but for those in the flight and ground training sector, there are unique challenges and untapped opportunities that require a nuanced approach. With content that stretches from search engine optimization and social media marketing to email campaigns, this guide brings together industry-specific advice to provide a comprehensive marketing playbook.

If you're looking to fill your classrooms—both virtual and physical—with eager student pilots, generate buzz about your training programs, or even build a community around your aviation offerings, this book will be your co-pilot, as you navigate the rapidly evolving digital marketing landscape. With real-world case studies, checklists, and expert opinions, we cover the A to Z of digital marketing tailored to the specific needs and nuances of aviation training. So, fasten your seatbelts and prepare for takeoff.

Shane D Jordan

> “THE AIM OF MARKETING IS TO KNOW AND UNDERSTAND THE CUSTOMER SO WELL THE PRODUCT OR SERVICE FITS HIM AND SELLS ITSELF.”

PETER DRUCKER

CONTENTS

1

TODAY'S FLIGHT STUDENTS ARE DIGITAL NATIVES, EXPECTING INSTANT INFORMATION AND SEAMLESS INTERACTIONS

CHAPTER ONE

INTRODUCTION

THE IMPORTANCE OF DIGITAL MARKETING IN THE AVIATION TRAINING INDUSTRY.

Welcome to the first chapter of your definitive guide to mastering digital marketing for flight and ground schools. If you're reading this, you already recognize the importance of a strong digital presence in the aviation training industry. Yet, understanding its importance and effectively leveraging it are two very different things.

In this chapter, we'll explore why digital marketing is no longer optional for flight and ground schools, it's mission critical. We'll also outline how a focused digital marketing strategy can help you attract and retain more students, and ultimately, improve your profitability.

The aviation training industry has long relied on traditional marketing methods like print advertising, and direct mail campaigns. While these have their merits, the digital age has changed the way people search for and consume information. Today's students are digital natives. They expect instant information, personalized experiences, and seamless interactions across various platforms.

INTRODUCTION

THE EVOLVING LANDSCAPE OF AVIATION TRAINING

97% OF YOUNG ADULTS IN THE U.S. USE THE INTERNET.

92% CARRY A SMARTPHONE.

As of 2021, 93% of US adults use the Internet and that number is growing, according to Pew Research.[1] Prospective students are online looking for reviews, videos, and course content, and engaging with schools well before they ever step foot on an airstrip for a discovery flight or in a simulator.

The landscape of shopping has undergone a seismic shift over the past decade, moving away from purely brick-and-mortar experiences to a hybrid model that blends physical and digital touchpoints. A critical catalyst for this change has been the internet and the explosion of e-commerce. According to the U.S. Department of Commerce, e-commerce sales grew from $167.3 billion in 2010 to an astounding $277.6 billion in Q2 2023.[2] This impressive growth isn't just a result of people buying online; it's a reflection of how online platforms have made themselves an integral part of the entire shopping journey—from research to purchase—through an integrated and personalized customer experience.

Online research has become an inseparable component of modern consumer behavior, fundamentally transforming the pre-purchase phase. A Bazaar Voice survey found that, 89% of consumers research ratings and reviews online before making a purchase.[3] This isn't limited to high-ticket items like cars, electronics, or planes. Even for everyday low-cost products like groceries or cleaning supplies, consumers are increasingly turning to the web to read reviews, compare prices, and seek recommendations. Platforms like Amazon, Google Reviews, and specialized review sites like PilotTrainingReviews.com have emerged as critical avenues for consumer education. The trend is so pervasive that businesses across sectors are investing heavily in digital marketing, search engine optimization (SEO), and online customer engagement platforms, recognizing that their online presence can make or break a sale.

The power of online research extends to in-store shopping as well. A report by Salesforce revealed that 83% of U.S. shoppers used their phones during in-store shopping experiences to look up product information, read reviews, and compare prices.[4] Dubbed "ROPO" (Research Online, Purchase Offline), this behavior further highlights the interweaving of online and offline shopping experiences. For instance, a consumer might come across a product on social media, read reviews about it on a dedicated website, check it out in a physical store, and then either make the purchase in-store or go back online to find the best deal. It's a complex, multi-layered journey that leverages digital and physical channels.

Online research has also created a more educated and, consequently, more demanding consumer. The wealth of information available at their fingertips means that customers are not only better informed but also have higher expectations. In response, businesses, are adopting customer-centric approaches that prioritize personalized experiences, transparency, and quality assurance to meet these heightened expectations.

Based on an Accenture report, 91% of consumers are more likely to shop with brands that provide personalized offers and recommendations.[5] Companies that invest in delivering customized shopping experiences fueled by data analytics and AI algorithms, are better positioned to satisfy the informed consumer.

By adapting proven digital marketing tactics and e-commerce strategies from sectors like consumer retail, flight and ground schools can achieve more than just immediate lead conversions; they set the stage for a more competitive future by building a cycle of happy students that promote your school for you. These practices have evolved beyond mere trends; they have become ingrained in consumer expectations. As you continue reading, we'll teach you how to build a marketing plan designed to capitalize on these opportunities, setting you ahead of your competitors.

As the digital age progresses, the amalgamation of online research and shopping is set to become even more intricate and indispensable. With trends such as virtual try-ons, AI-based personal shopping assistants, and augmented reality product previews on the horizon, the line between online research and actual purchasing is becoming increasingly blurred, heralding an even more interconnected and digital-first retail landscape.

By integrating online research into every phase of the shopping journey, both consumers and businesses are experiencing a shift that is not just a change but an evolution—where data-driven insights and personalized experiences rule the day.

> **"INNOVATION NEEDS TO BE PART OF YOUR CULTURE. CONSUMERS ARE TRANSFORMING FASTER THAN WE ARE, AND IF WE DON'T CATCH UP, WE'RE IN TROUBLE."**
>
> Ian Schafer

WHY DIGITAL MARKETING IS A GAME-CHANGER

Targeted Audiences

In today's world, old-school marketing methods are getting left behind as digital ways of reaching people take over. Flight schools, which train pilots, are a perfect example of how digital marketing can make a big difference. The main reason? Digital marketing lets schools talk directly to the people they want to reach.

By using digital marketing, flight schools can really get to know what potential student pilots want. Instead of casting a wide net and hoping for the best, they can use data to see if students are looking for certain types of training, flexible class times, or modern equipment. By matching their offers to these needs, schools can make sure their ads and messages really connect with people.

Another big plus is that digital marketing can save money. Instead of spending a lot on ads that might not hit the mark, schools can use their budget more wisely. They can create ads and messages that speak directly to the people they want to attract. This means they get more students for less money.

Building trust is another big part of why digital marketing works so well. Schools can share useful articles or stories about flight training, safety tips, or success stories. This shows they know their stuff, which can help bring in new students and keep them loyal for the long run.

What's also cool about digital marketing is that it can reach people at just the right time. Maybe someone is just starting to think about becoming a pilot, or maybe they're already looking for a good flight school. Either way, the right ads and content can guide them to make a decision. By being there at the right moment, flight schools can make sure they're the first choice for new students.

Today's students also like things that feel personal. With digital marketing, flight schools can send messages or offers that fit what each person is looking for. This makes students feel understood and can help convince them to sign up.

Lastly, one of the best things about digital marketing is that schools can see how well their ads are working right away. They can check if people are clicking on their ads, reading their content, or signing up for classes. If something's not working, they can change it up and try again. This means they can always be improving and doing their best.

In short, as the world of flying and pilot training changes, flight schools need to keep up. Using digital marketing to reach the right people at the right time is a smart move. It can help schools get more students, keep them happy, and grow bigger and better. For flight schools that want to stand out and succeed, digital marketing is the way to go.

Wider Reach

The beauty of digital is its unlimited, cost-effective reach. Where flyers and billboards are restricted by geography, your digital ads can span a nation and even the world. According to Google, businesses make an average of $2 in income for every $1 they spend on Google Ads.[6] This return on investment (ROI) is almost unimaginable in traditional advertising. This fact is incredibly pertinent for flight schools, an industry that has traditionally relied on local, often outdated, marketing methods to attract new students. Let's be clear: the old-school "Learn to Fly Here" signs staked out by the roadside or posted at the local airfield simply won't cut it anymore.

> **TRADITIONAL ADVERTISING IS A MONOLOGUE, BUT DIGITAL ADVERTISING IS A DIALOGUE.**

The target market for flight schools is has grown much larger than the community within a 50-mile radius. People are willing to move or travel significant distances to attend a high-quality training program. This means your potential reach is effectively national, if not international. Imagine someone in a distant state, or another country, searching for "best flight schools" or seeing your Google Ad, and landing on your website. Your scope for attracting students has just increased exponentially, breaking the boundaries traditional advertising could never breach.

YOUR DIGITAL PLATFORMS AREN'T JUST BULLETIN BOARDS; **THEY'RE DIALOGUE SPACES.**

Digital also offers a richer palette for storytelling. Traditional signs with their limited text and visual space can't convey the depth of experience your flight school offers—the state-of-the-art simulators, the expert instructors, or the success stories of your alumni. A well-designed digital ad can link to a landing page filled with video testimonials, virtual tours, and detailed course information. You could even implement chatbots to answer frequently asked questions, providing instant responses and establishing an immediate connection with potential students.

The data analytics available with digital advertising give flight schools unprecedented insights into their audience. You can learn about the demographics of the people most interested in your programs, what keywords they're using in their searches, and which elements of your advertising are most effective. This information is invaluable for refining your marketing strategies. In contrast, traditional advertising is often a shot in the dark—you don't really know who's seeing your signs and whether they have any impact.

Additionally, online platforms allow for a two-way conversation. Traditional advertising is a monologue, but digital advertising is a dialogue. When someone clicks on your ad, they can immediately engage with your content through liking, sharing, commenting, or even direct messaging. This interaction opens the door to relationship-building, which is crucial for converting interest into enrollment.

Finally, let's talk about scalability and adaptability. Unlike physical signs or billboards, which require a significant investment of time and money to update, digital ads can be tweaked in real-time to test different messages, images, or calls-to-action. You're not locked into a single, static message; you can adapt and evolve as you learn more about what resonates with your target audience.

The marketing landscape has shifted, and flight schools can no longer rely solely on traditional means of advertising and expect to remain competitive. Digital marketing offers reach, richness, and an ROI that "Learn to Fly Here" signs can't match. The sooner flight schools recognize this and adapt, the better positioned they'll be in a marketplace that has already taken to the skies.

Higher Engagement

Let's amp up the conversation. Your digital platforms aren't just bulletin boards; they have the potential to create meaningful interactions. Want to master the art of two-way conversation? Chapter 8: "Social Media Marketing" is your go-to resource.

As HubSpot reported, companies that prioritize customer engagement strategies see 1.7 times faster revenue growth than competitors who don't.[7] That's a critical stat, considering how saturated the market can be for specialized industries like aviation training schools.

It's important to point out that engagement isn't just a vanity metric; it's often directly correlated with customer loyalty and, ultimately, revenue. A study by Harvard Business Review (HBR) revealed that customers who engage with a brand online are more likely to visit that business offline.[8] Imagine the exponential potential for aviation schools here. Engaging with prospective pilots online increases your chances of converting them into enrolled students who'll walk through your doors, ready to fly.

DATA IS THE NEW FUEL, **DRIVING DECISION-MAKING WITH NUMBERS** RATHER THAN GUT FEELINGS.

Social media plays an instrumental role in amplifying these interactions and establishing brand authority. Platforms like LinkedIn, Instagram,

ACCORDING TO A REPORT BY GOOGLE, COMPANIES USING GOOGLE ANALYTICS HAVE FOUND THAT THEY CAN REDUCE THEIR MARKETING SPEND BY AS MUCH AS 25% WHILE STILL ACHIEVING THE SAME PERFORMANCE.

TikTok, and X (formerly known as Twitter) allow you to share valuable content that not only promotes your training programs but also positions you as an industry thought leader. For flight schools, this could mean anything from sharing student testimonials, pilot tips, or industry news to hosting live Q&A sessions on Facebook or Instagram to answer common queries about pilot training. Building brand authority on these channels is proven to drive conversions. According to Forbes, 76% of social media users have purchased something they saw on a social media platform.[9]

LEVERAGE ANALYTICS TO **OUTPERFORM YOUR COMPETITORS** BY **23x**

Digital marketing channels allow you to build long-term relationships with your audience. By offering value through educational content, you're not just selling a service; you're becoming a trusted advisor. Think about writing a blog that answers FAQs about aviation training or producing webinars that walk prospective students through the enrollment process or showcase career opportunities post-training. Over time, these content pieces not only improve your search engine rankings but also make you the go-to resource for anything related to aviation training.

Engagement isn't just a buzzword; it's a robust, multidimensional strategy that incorporates customer interaction, content marketing, and social media prowess to enhance your brand's authority and drive sales. All of these facets are interconnected and essential for optimizing your digital marketing to its fullest potential. With proper engagement, you're not just broadcasting; you're conversing, educating, and building lasting relationships.

Data-Driven Decisions

Data is the new fuel, driving decision-making with numbers rather than gut feelings. Analytics will be your best friend, and we explore this in depth in "Part IV: Analytics and ROI." The application of analytics in your digital marketing strategy can be a game-changer, especially in terms of ROI. A study by McKinsey found organizations that leverage customer analytics are 23 times more likely to outperform competitors in new customer acquisition and six times more likely to retain customers.[10] Moreover, data-driven companies are 19 times more likely to achieve above-average profitability.

Let's talk about optimization. With the right set of analytics, you can precisely measure what's working and what's not in your marketing campaigns. This allows for real-time adjustments, which are often impossible with traditional advertising means. For example, if you see

THE BEAUTIFUL PART ABOUT DIGITAL MARKETING IS ITS SCALABILITY.

that a specific keyword is underperforming in a Google Ads campaign for your flight school, you can immediately pause it and allocate that budget to a better-performing keyword.

Advanced analytics can offer predictive insights, not just descriptive ones. This means you can forecast trends and consumer behaviors based on the data you've gathered, allowing you to make proactive decisions that can potentially save money while increasing ROI. Forrester research indicates that predictive analytics can improve ROI by up to 3.2 times when properly implemented.[11] This can be invaluable in long-term planning and can guide the allocation of resources more efficiently.

WITHOUT ANALYTICS, YOU ARE FLYING BLIND IN A LANDSCAPE THAT DEMANDS PRECISION NAVIGATION.

In a specialized industry like aviation training, analytics can help you effectively and efficiently identify your core audience. Demographics, behaviors, interests, and geographic location are just some of the parameters you can leverage to target potential students who are most likely to enroll in your programs, maximizing your advertising ROI.

It's not an overstatement to say that analytics equip you with the tools to make your marketing spend more efficient, targeted, and, ultimately, more profitable. Businesses that shy away from analytics are essentially flying blind in a landscape that demands precision navigation. By adopting analytics, you're not only conserving resources but also setting your business on a trajectory toward higher profitability and sustainability.

Cost-Effectiveness

Are you looking to stretch every dollar in your budget? Digital marketing is the financial advisor you didn't know you needed. How to get the most bang for your buck will be covered in Chapter 10: "Budgeting and ROI." The financial implications of a well-executed digital marketing strategy can be monumental, especially for small businesses like flight schools, which typically operate on tighter budgets. According to a report by HubSpot,

PROVIDING ADDITIONAL DIGITAL SERVICES EXISTING PRODUCT OFFERINGS **CAN INCREASE REVENUE BY UP TO 35%**

digital marketing generates a cost per lead that is 61% less expensive than traditional marketing channels. This cost-efficiency enables you to allocate resources more effectively and to other operational areas that require attention.

Furthermore, it's not just about the upfront cost per lead; digital marketing also influences the lifetime value of your customer. A study by Adobe found that customers acquired through search engine marketing and social media have a 21% higher lifetime value compared to those acquired through other channels. This is particularly critical for aviation training centers where the customer journey is long-term and highly involved. By leveraging digital marketing to its fullest potential, you can not only attract quality leads but also foster long-lasting relationships that translate into greater Customer Lifetime Value (CLV) which we'll explore in Chapter 19.

Another essential metric to consider is Return on Ad Spend (ROAS), a measure of gross revenue generated for every dollar spent on advertising. A WordStream study indicates that the average ROAS across all industries is $2:1, but with effective targeting and analytics, businesses can achieve ROAS as high as $8:1. In specialized sectors like aviation training, where

OVERALL, THE POTENTIAL FOR PROFIT THROUGH DIGITAL MONETIZATION IS NOT JUST HYPOTHETICAL; IT'S BACKED BY HARD DATA AND REAL-WORLD EXAMPLES. BY EXPLORING THESE VARIOUS CHANNELS, YOU CAN DIVERSIFY YOUR REVENUE STREAMS AND SIGNIFICANTLY ENHANCE YOUR PROFIT MARGINS.

the customer lifetime value can be substantial, the ROAS potential can be even more rewarding, making every dollar spent even more effective.

The beautiful part about digital marketing is its scalability. Even a small business can start with a modest budget and scale up as you see the results. The immediacy of analytics allows you to see what works in real-time and optimize accordingly without having to wait for long-term campaign results. According to the Small Business Administration, businesses that adapt digital marketing can save 20-30% in marketing costs while increasing revenue by up to 30%.

Digital marketing offers a pathway to more efficient financial management of your advertising budget. Its ability to generate higher ROI, reduce cost per lead, and maximize lifetime customer value makes it an indispensable tool for small businesses. While traditional marketing methods have their place, the future—and most of the present—belongs to digital.

WITHOUT ANALYTICS, YOU ARE FLYING BLIND IN A LANDSCAPE THAT DEMANDS PRECISION NAVIGATION.

Monetization Opportunities

Sponsored posts, affiliate marketing, premium content—you name it, you can monetize it. For a step-by-step guide on revenue generation through digital channels, make sure to read Chapter 11: "Monetization Strategies for the Digital Age." The monetization opportunities presented by digital marketing are vast and varied, and their profit potential is similarly robust. The affiliate marketing industry alone is worth $17 billion, and it's growing rapidly, showcasing the immense revenue potential for those willing to invest in digital channels.[12] This is significant for niche businesses like aviation training companies, as they can partner with relevant brands to generate additional income.

Sponsored posts also offer a lucrative avenue for revenue generation. According to Insider Intelligence, influencer marketing spend in the US will exceed $2 billion this year.[13] By producing high-quality, informative content and aligning with brands that share your target audience, you can tap into this multi-billion-dollar market. In fact, premium content is another revenue stream to consider. A Deloitte report discovered that consumers are willing to pay a premium for exclusive digital content. This marks a tremendous opportunity to create and share thought leadership content and get paid for doing so!

Don't forget about monetizing data. While often overlooked, data monetization can generate significant profits. McKinsey estimates that data monetization could become a $300-billion annual revenue opportunity. For example, aviation training centers could potentially sell anonymized data on pilot training trends, equipment preferences, or career progression to interested parties like aviation manufacturers or industry researchers.

Moreover, let's talk about value-added services. Forrester found that businesses that provide additional digital services to their existing product offerings can increase their revenue by up to 35%. For instance, you can offer exclusive webinars, advanced tutorials, or even one-on-one coaching sessions for an additional fee. These specialized services not only contribute directly to your bottom line but also increase customer engagement and lifetime value.

Overall, the potential for profit through digital monetization is not just hypothetical; it's backed by hard data and real-world examples. By exploring these options, you can diversify your revenue streams and significantly enhance your profit margins.

ALIGNING DIGITAL MARKETING WITH BUSINESS GOALS

The digital realm isn't a playground; it's a strategic board game. Your moves need to connect to your objectives. For a detailed plan on setting and achieving business goals through digital marketing, stay tuned for Chapter 12: "Aligning Business Goals with Digital Strategy." When your digital marketing initiatives are in perfect alignment with your business goals, a powerful synergy is created that amplifies your efforts and propels you towards success. A report by the Content Marketing Institute states that businesses with a documented strategy are 313% more likely to report success in their marketing endeavors compared to those without one. In essence, alignment is not an option; it's a necessity for achieving optimal performance.

Alignment facilitates resource optimization as well. Organizations with highly synchronized marketing and business objectives enjoy a 36% higher customer retention rate and a 38% higher sales win rate, per an Aberdeen Group Study. By integrating your digital strategy with your business objectives, you ensure that every dollar spent on digital marketing directly contributes to organizational goals such as revenue growth, customer acquisition, or market expansion.

A report from McKinsey also underscores the financial implications of this alignment. It indicates that companies with closely aligned marketing and business strategies report up to 30% higher shareholder returns. This can be particularly impactful for smaller and more specialized businesses like flight schools or aviation training centers, where even minor adjustments in strategy can lead to significant financial gains.

Let's consider customer engagement, a key performance indicator for many businesses. Adobe's Digital Intelligence Briefing states that companies that align their digital marketing strategy with customer engagement goals see a 20% increase in customer satisfaction rates. In fields like aviation training, this could translate into higher enrollment rates and more positive reviews—

indicators that serve your immediate financial goals and build long-term brand equity.

Additionally, alignment allows for better measurement and analytics. An MIT Sloan study suggests that companies with aligned strategies are 2.2 times more likely to say that their data and analytics efforts have produced a significant business impact. By understanding how your digital marketing activities are driving progress towards business objectives, you can adjust your strategy dynamically, ensuring that you're always on the most effective and efficient path to achieving your goals.

In summary, alignment between your digital marketing and broader business strategy doesn't just create a powerful synergy; it can positively impact multiple business metrics, including financial performance, customer retention, or operational efficiency.

PREPARING FOR THE JOURNEY AHEAD

Digital marketing isn't optional—it's imperative. If you're not actively promoting your business online, you're basically invisible to a significant portion of your potential customers—and you're leaving revenue potential untapped.

What's next? Identifying your target audience. Who are they, what do they want, and how can you give it to them? When you understand who you're marketing to, you'll be able to create a highly targeted, effective strategy that resonates with your audience and delivers measurable results.

READY TO LAUNCH? AS WE TAXI DOWN THE RUNWAY, REMEMBER:

This is just the beginning. Each subsequent chapter offers in depth knowledge, tools, and actionable insights tailored to the aviation training industry. Keep that seatbelt fastened; it's going to be an exhilarating ride!

PART I

STRATEGY

2

DEFINE YOUR AUDIENCE AND CONVERT 2X MORE LEADS THAN YOUR COMPETITORS

CHAPTER TWO

DEFINE YOUR AUDIENCE

UNDERSTANDING YOUR TARGET AUDIENCE: STUDENT PILOTS AND AVIATION PROFESSIONALS

Your marketing efforts, no matter how elaborate or creative, will fall flat if they don't speak directly to the needs and interests of your target audience. The aviation training sector has a unique blend of student pilots and aviation professionals requiring specific services and training. To appeal to them, you'll need more than generic advertising; you'll need to craft messages that resonate on a personal and professional level.

According to Forbes, businesses that tailor their marketing to a well-defined target audience are twice as likely to convert a lead into a sale. This chapter will arm you with actionable insights to define and connect with your target audience effectively, so you can turn those potentials into pilots.

We'll now work together to define your audience and show you how to change your marketing strategy based on that group's preferences and habits.

AUDIENCE

IDENTIFYING PERSONAS

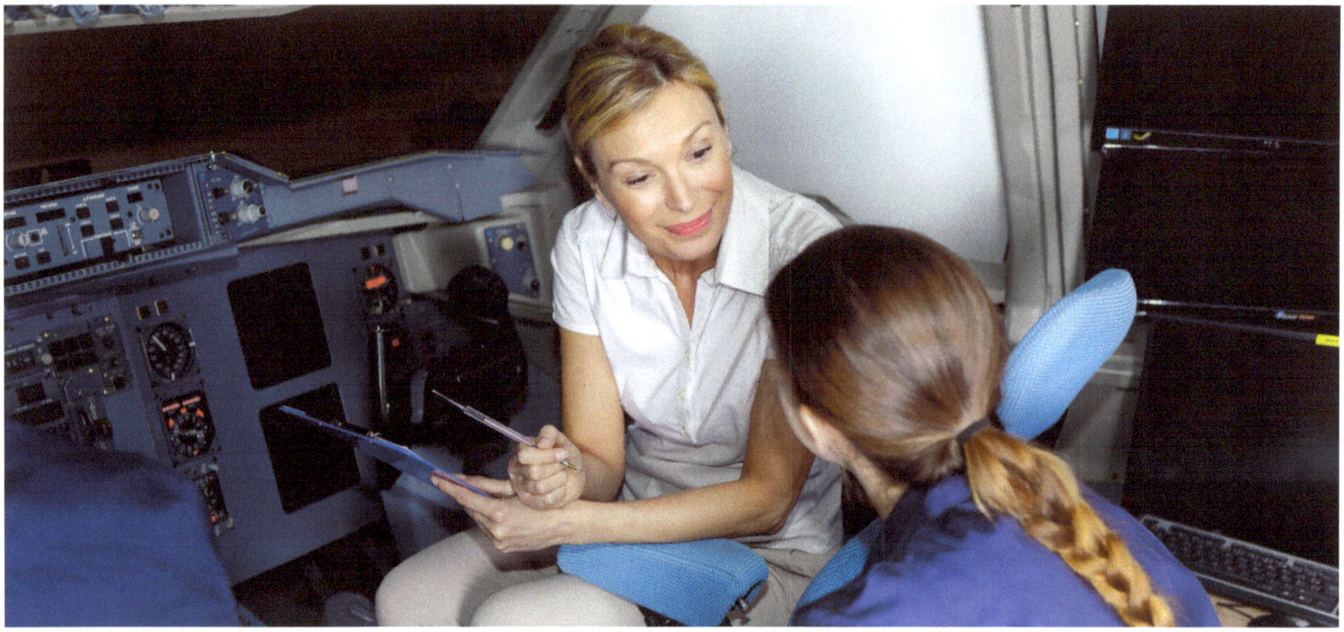

IT'S NOT JUST ABOUT DEMOGRAPHICS, **BUT PSYCHO-GRAPHICS AS WELL.**

Creating target personas is not just a trendy marketing gimmick; it's a strategic imperative, particularly for small businesses like aviation training centers. The concept of a persona is akin to a detailed character sketch representing a specific audience segment. Each persona should have a name, age, job title, pain points, and aspirations related to aviation training. While larger companies can sometimes afford a scattershot approach to marketing—casting a wide net to see what they can catch—small businesses need to be more targeted to make the most out of their limited marketing budgets.

We cannot overstate the importance of building accurate customer personas. According to a study published by IBM's Institute for Business Value, companies that excel at using customer personas in their marketing strategies see an 83% increase in performance metrics such as conversion rates and lead generation. By establishing these detailed profiles, you're developing a deeper understanding of who your customer is, what motivates them, and how best to reach them.

It's not just about demographics, but psychographics as well. While demographics—such as age and job title—help to paint a broad picture, psychographics dive deep into attitudes, emotions, and the specific pain points that prospective students or professionals in aviation may have. Is

PERSONA 1

Sarah, a 25-year-old mechanical engineer interested in becoming a commercial pilot, but apprehensive about the financial investment.

PERSONA 2

Michael, a 35-year-old aviation mechanic looking for advanced certification.

cost a potential barrier? Are they looking for flexible training schedules? These details help tailor your marketing message for maximum resonance.

For small businesses, each customer holds significantly more weight in terms of lifetime value and word-of-mouth referrals. Small Business Trends reports that 85% of small businesses get customers through word of mouth. By accurately identifying your customers, you can craft personalized, effective messaging that not only attracts but also delights your customers, turning them into advocates for your business.

Creating customer personas also leads to more efficient use of marketing resources. By eliminating wasted marketing efforts that engage the wrong audience. According to Marketing Insider Group, businesses see a 73% increase in marketing efficiency when they use customer personas to guide their content and strategy decisions. This increased effectiveness allows small businesses to compete effectively with more prominent players in the industry by targeting niche markets with specialized needs—like aspiring pilots or aviation mechanics looking for advanced training that larger companies might overlook.

CUSTOMER PERSONAS ARE NOT A MARKETING LUXURY **BUT A NECESSITY.**

With well-defined personas, you'll be able to promote the right training services to the right people at the right time, thereby ensuring higher engagement, conversion rates, and long-term customer relationships.

MARKET RESEARCH

Market research is the backbone of any successful marketing strategy, and it's particularly crucial for gathering information to create effective customer personas. Tools like customer interviews, surveys, and data analytics offer invaluable insights into your potential clients, what they are looking for, and how your flight school can address their needs. It's not an area skim over. Harvard Business Review found that companies that invest in creating customer personas based on comprehensive market research are 60% more likely to exceed their revenue and lead goals. This statistic provides compelling evidence that market research isn't just helpful; it's essential.

42% OF SMALL BUSINESSES **FAIL DUE TO A LACK** OF MARKET DEMAND.

For small businesses, market research holds even more significance. Unlike large corporations with extensive budgets and resources, small businesses often operate under limitations and can't afford to make costly mistakes. A study by CBInsights revealed that 42% of small businesses fail due to a lack of market demand. Conducting rigorous market research helps de-risk this scenario, providing insights into market trends, customer behavior, and competitive analysis, thereby helping businesses align their offerings to actual market needs.

Customer interviews are a valuable resource, allowing for deep dives into their experiences, pain points, and expectations. You get qualitative data, filled with narratives and stories that numbers alone can't tell. Surveys, on the other hand, provide quantitative data. They can be distributed widely and can quickly accumulate a large volume of responses that can be statistically analyzed. According to the Small Business Administration, businesses that conduct regular surveys have a 31% higher customer retention rate.

Data analytics is the third pillar, offering insights that are not immediately visible. Using analytics tools, you can track customer behaviors, identify patterns, and understand preferences. Even basic website analytics can offer a wealth of information. For instance, Google Analytics can show you how long users stay on your site, which pages they visit, and what content engages them most. Businesses that leverage data analytics have been shown to have a 23% better chance of outperforming their competitors in terms of customer satisfaction, per a McKinsey report.

As you gather this information, the outlines of your personas will become sharper, enabling you to craft more targeted and effective marketing strategies. By meticulously segmenting your audience, you can create highly personalized experiences. This level of personalization often translates into higher conversion rates, more significant customer loyalty, and, ultimately, increased profitability, particularly for small businesses that need to make every interaction count.

ACTION STEPS

1. Conduct surveys asking your current students or interested parties about their goals, challenges, and demographics.
2. Analyze website and social media analytics to determine the characteristics of your most engaged audience.

DATA INTERPRETATION

The process of collating your market research data into relevant categories—such as age, career aspirations, or financial readiness—is an indispensable step in shaping a precise and effective marketing strategy. This categorization is an analytical approach that can yield actionable insights into customer behavior, preferences, and expectations. By doing so, you can identify patterns and common pain points that will directly influence and refine your marketing strategies, thus optimizing your chances for business success.

According to Forbes, 64% of marketers see data-driven strategies as vital in today's hyper-competitive economy. The importance of this step is further validated by a McKinsey study showing that companies who put data at the center of their marketing and sales decisions improve their marketing ROI by 15-20%. The essence here is not just collecting raw data but converting it into organized, interpretable information that can guide decision-making processes. Businesses that effectively categorize their customer data are better positioned to achieve their objectives, as they can tailor their services and marketing campaigns to meet the unique needs and preferences of different customer segments.

Categorizing data enables a nuanced understanding of customer journeys, allowing businesses to identify and rectify gaps in their service delivery and marketing funnels. A study by Adobe shows that companies that invest in a holistic understanding of the customer journey experience a 25% increase in their close rate. By understanding your customer's common pain points, you can craft compelling marketing messages that directly address them, increasing relevance and consumer engagement.

Data categorization allows small businesses to concentrate their efforts on more targeted, high-ROI activities. Gartner's CMO Spend Survey indicates almost 80% of small businesses that focused on data-driven personalized marketing saw an increase in customer engagement levels within three months.

For an industry like aviation training, this specificity can be a game-changer. Knowing your potential customer's career aspirations or financial readiness can provide valuable context. For example, if the majority of your potential students are career switchers who are financially secure, your messaging may focus more on the quality of training and the opportunities it opens, rather than affordability.

The process of categorizing your data is far from a clerical chore; it is an essential component of a data-driven marketing strategy that has proven to yield superior ROI and customer engagement levels. By implementing this step meticulously, you pave the way for personalized, effective, and successful marketing campaigns.

TAILORING YOUR MARKETING STRATEGY

In an increasingly competitive marketplace, a one-size-fits-all approach to marketing is no longer effective or sustainable. Tailoring your strategy to meet the unique needs, preferences, and pain points of your target audience is crucial for standing out and achieving measurable results. A bespoke marketing strategy not only enhances engagement and customer loyalty but also optimizes your resources for higher ROI. In essence, personalized marketing is a necessity for any business aiming for long-term success and profitability.

Personalization

We're living in an era where personalization has become a fundamental consumer expectation. Salesforce discovered that for a staggering 84% of customers being treated as a unique individual, rather than just another sales figure, is key to winning their business. This isn't a negligible percentage; it's a loud and clear message businesses can't afford to ignore.

The need for personalization extends beyond customer service; it infiltrates every aspect of the customer journey, from initial awareness and consideration through to purchase and post-purchase engagement. Personalization is a powerful business strategy that has a direct impact on the bottom line. Brands that create personalized experiences are seeing revenue increases of 6% to 10%—two to three times faster than those that don't—according to a study by Boston Consulting Group (BCG).

Personalization doesn't just improve revenue; it enhances customer loyalty, a vital factor for long-term success. A report by Accenture shows that 91% of consumers are more likely to shop with brands who recognize, remember, and provide relevant offers and recommendations. Loyalty is particularly important in the digital age, where customers have an abundance of options at their fingertips. Retaining a customer is often more cost-effective than acquiring a new one. Acquiring a new customer is five to 25 times more expensive than retaining an existing one, and increasing customer retention rates by 5% increases profits by 25% to 95%, per a study by HBR.

Another aspect worth mentioning is the positive impact of personalization on your marketing ROI. A study by McKinsey & Company highlights that personalization can deliver 5-8 times the ROI on marketing spend, and can lift sales by 10% or more. This degree of effectiveness is invaluable for small and medium-sized businesses with limited marketing budgets.

However, personalization isn't just about plastering the customer's name all over your e-newsletters. It involves data-driven strategies that consider customer behaviors, preferences, and previous interactions with your brand. The use of Artificial Intelligence and Machine Learning in crafting personalized experiences is rising, and the results are noteworthy. According to Capgemini, 75% of organizations implementing AI and machine learning for personalization have increased customer satisfaction scores by at least 10%.

Personalization equates to relevancy, and relevancy increases to engagement. As technology advances and data analytics become more sophisticated, the potential for deeply personalized customer experiences will only expand. Ignoring this crucial trend risks not just stagnation, but potential business decline.

Content Marketing

The maxim "Content is king" is often thrown around in the marketing world, but its weight has never been more substantial than it is today. In an age of information overload, merely having content isn't enough—it has to be exceptional, useful, and most importantly, tailored to your specific audience. Adding context to your content is what converts it from mere information to a targeted solution. According to a Content Marketing Institute study, companies that are leaders in content marketing experience 7.8 times more site traffic compared to those who don't invest as much in content. That's not just a statistical blip; it's a testament to the power of high-quality, relevant content.

The significance of content marketing extends well beyond traffic generation; it plays a vital role in customer engagement, brand perception, and ultimately, sales conversion. According to HubSpot, companies that publish more than 16 blog posts a month get nearly 3.5 times more traffic and 4.5 times more leads than companies that publish four or fewer monthly posts. This dramatically increases opportunities for conversion and thereby boosts the overall ROI.

ACTION STEPS

1. Implement AI chatbots on your website to provide personalized course recommendations based on individual queries.
2. Use segmented email lists to send out targeted promotions or information.

A properly executed content marketing strategy can significantly enhance your brand's authority and trustworthiness. By offering valuable insights, solutions, and thought leadership, you're not just selling a product or service; you're building a relationship. Demand Gen Report discovered that, 47% of buyers viewed three to five pieces of content before engaging with a sales rep. Furthermore, 96% of the most successful content marketers agreed that their audience views their organization as a credible and trusted resource, as per CMI's B2B content marketing report.

And let's talk numbers. A study by Kapost found that content marketing brings in three times as many leads as traditional marketing and costs 62% less. When you consider the lifetime value of a customer, the ROI of content marketing can be colossal. Inbound marketing attracts customers through relevant content and engagement, while outbound marketing directly promotes products or services to a broad audience, often through traditional advertising methods. According to a report by MarketingProfs, content marketing costs 41% less per lead for companies focusing primarily on inbound marketing compared to those who invest mainly in outbound marketing strategies.

CONTENT MARKETING BRINGS IN THREE TIMES AS MANY LEADS AS TRADITIONAL MARKETING AND COSTS **62%** LESS.

Moreover, content marketing integrates seamlessly with other digital marketing strategies like SEO, paid search and retargeting, creating a synergistic effect that amplifies your reach and engagement. A study by Gartner revealed that integrated campaigns across four or more digital channels outperform single or dual-channel campaigns by 300%. This suggests that when your content marketing is synchronized with your overall digital marketing strategy, your ROI multiplies.

Content marketing is a cornerstone of modern marketing—integral to brand building, lead generation, and sales conversion. By aligning your content to the specific problems, needs, and interests of your customer personas, you're not just throwing a dart in the dark; you're making a calculated, data-backed move to capture and retain your target audience.

Social Media Platforms

Over the past 25 years, social media has evolved from a novel way to connect with friends and family, to a legitimate and important marketing tool. Now, it's a pivotal channel not just for brand awareness but for sales, customer service, and reputation management as well. Statista estimates there are 4.89 billion social media users worldwide, projects that will increase to 5.85 users by 2027.[14] Ignoring such a broad audience could be detrimental to a business aiming for scalable growth.

There are numerous social platforms that serve different purposes and cater to various demographics, as highlighted by the Pew Research study. Understanding your target audience is crucial for choosing the appropriate social media channels for your business. For example, if your target persona is a young adult interested in aviation, Instagram's visual-heavy format and younger demographic (71% of its users are under the age of 35, according to Sprout Social) make it an excellent fit. In contrast, if you're targeting aviation professionals for advanced training programs, LinkedIn, with its professional user base, would be more suitable.

It's not just about selecting the right platform; it's also about how you leverage it. Paid social media offers the kind of targeting and customization options that are unheard of in traditional media. For instance, Facebook's ad platform allows businesses to focus on users based on location, age, likes, interests, and behavior. An eMarketer report reveals worldwide spending on social media advertising is expected to reach $110 billion by 2022, with ROI surpassing traditional advertising channels. Moreover, Adobe states that social media has a 100% higher lead-to-close rate than outbound marketing.

One of the most beneficial aspects of ROI from social media is customer engagement. A study from J.D. Power found that 67% of consumers have used a company's social media channel for customer service. And when they do, they expect a fast response. According to Sprout Social, 40% of consumers expect a response within an hour. Quick, effective customer service on social media leads to higher customer satisfaction and amplifies the likelihood of repeat business and positive reviews. Timely responses can quickly resolve customer service issues before they escalate, protecting your brand from potentially negative comments on your social channels.

ACTION STEPS

1. Determine which social media platforms are most most used by your target audience.
2. Create a social strategy that provides a balance of educational and promotional content.
3. Develop a posting schedule that aligns with when your target audience is most active.

IT'S NOT JUST ABOUT SELECTING THE RIGHT PLATFORM; IT'S ALSO ABOUT **HOW YOU LEVERAGE IT.**

Social media plays a significant role in content distribution and SEO, both of which are vital for inbound marketing. According to Social Media Examiner's annual report, 58% of marketers said that social media has helped them improve sales. Also, Google has started to include tweets in their search queries, highlighting the growing importance of social media in search rankings.

The native social media platforms provide a myriad of analytics that give clear insights into engagement rates, reach, likes, shares, and conversions. According to Buffer, businesses that utilize social analytics see a 126% higher median ROI than those who don't. This data-driven approach enables constant refinement of your marketing strategies, making social media a critical business intelligence asset for your school.

Social media isn't just an add-on to your digital marketing strategy; it's a powerhouse in its own right. By choosing the right platforms based on your target demographics and employing a data-driven, highly tailored strategy, you can exponentially improve your brand's reach, engagement, and ROI.

ADAPTING TO CHANGE

The pace of change in today's marketplace is unlike anything we've seen in the past. Due to rapid technological advancements, evolving social trends, and shifts in consumer behavior, what worked yesterday might not work tomorrow. Businesses need to adapt their marketing strategies to stay relevant and profitable. A Gartner study highlights how fluid the consumer-brand relationship is: 77% of consumers' brand perceptions change continuously due to active evaluation or experience. This insight is a call to action to ensure your brand remains top-of-mind for these discerning consumers.

Given this dynamic landscape, it's crucial to incorporate flexibility and adaptability into your marketing strategy. Gartner's report also suggests that businesses should be prepared to pivot their marketing strategies rapidly to accommodate fluctuating consumer perceptions. Failing to adapt can not only result in decreased market share but also in a significant decrease in ROI. For example, NewVoiceMedia reports that poor customer service experiences cost businesses $75 billion a year.[15]

Now, let's talk about data-driven decisions. According to McKinsey & Company, businesses that make data-driven decisions are 19 times more likely to be profitable. Utilizing real-time analytics tools can give you actionable insights into consumer behaviors, preferences, and perceptions. These tools allow you to adapt your marketing strategies almost instantly. With data analytics, you can fine-tune your messaging, focus your advertising spend, and target different consumer segments more effectively. In other words, data doesn't just help you understand your current situation; it helps you predict your future.

In addition to consumer behavior, changes in technology also necessitate strategy shifts. 76% of participants in an Adobe study feel that marketing has changed more in the past two years than it did over the previous fifty. New channels, platforms, and devices appear continually. The surge in voice-activated technologies, for example, is reshaping the way brands think about SEO and online content. Companies that have adapted their content for voice search are already reaping the benefits. Demand Sage states that more than 50% of adults reported using voice search daily, which means ignoring this trend could result in a significant loss of online visibility for your brand.[16]

ACTION STEPS

1. Create blog posts and video content that directly address the challenges and questions your target audience has about aviation training.
2. Share testimonials or case studies that reflect similar journeys to those of your personas.

ACTION STEPS

1. Regularly review and update your personas.
2. Keep an eye on customer reviews and social media discussions to understand emerging trends or shifts in perception.

Market conditions can change rapidly due to various external factors such as economic downturns, global pandemics, or changes in regulations. According to HBR, companies that maintain their marketing spend during economic downturns tend to fare better in the long run compared to those who cut their budgets. This is because they are better positioned to capitalize on the eventual upturn and can gain market share from less forward-thinking competitors in the meantime.

Flexibility in your marketing strategy is required for the survival and growth of your flight school. Whether it's changing consumer behavior, advances in technology, or shifts in the economic landscape, your ability to adapt will define your long-term success. Your marketing strategy should be a living, breathing document, backed by data and able to pivot at a moment's notice. Such flexibility doesn't just help you navigate risks; it also uncovers new opportunities for better ROI.

CONCLUSION

Understanding your target audience is the cornerstone of effective digital marketing. In this high-stakes game of flying high or falling flat, the more tailored your approach, the more impactful your results will be. In the next chapter, we'll discuss the various digital marketing channels available to you and how you can leverage them to reach your well-defined target audience successfully.

Ready to dig deeper? Proceed to Chapter 3, where we'll delve into the different avenues you can use to connect with your target audience through digital marketing.

READY TO DIG DEEPER?

Proceed to Chapter 3, where we'll delve into the different avenues you can use to connect with your target audience through digital marketing.

3

DEFINE, CREATE, AND IMPLEMENT YOUR LONG-TERM DIGITAL MARKETING STRATEGY

CHAPTER THREE

DEFINE YOUR STRATEGY

CREATING A LONG-TERM DIGITAL MARKETING PLAN

A robust digital marketing strategy will be the compass that guides your flight school towards success. In this chapter, we'll explore the significance of a digital marketing strategy, its alignment with your business plan, the essential steps to create one, the critical influence of audience definition, and the importance of continuous strategy review and updates.

DIGITAL MARKETING PLAN

THE IMPORTANCE OF A DIGITAL MARKETING STRATEGY

In today's digital age, a well-crafted digital marketing strategy serves as the flight plan that ensures your marketing efforts are focused, effective, and aligned with your business objectives. Here's why it's crucial:

A STRATEGY SETS CLEAR GOALS AND KPIs ENABLE YOU TO **MEASURE THE IMPACT OF YOUR MARKETING EFFORTS ACCURATELY.**

1. Maximizing Online Visibility

Aspiring pilots and aviation enthusiasts increasingly rely on the internet to research and make decisions. A digital marketing strategy helps you capture their attention and ensure your flight school is highly visible online. Start with your website, see Chapter 4: Anatomy of a Successful Website. Go beyond website design with SEO in Chapter 5: Search Engine Optimization - Climbing the Ranks.

2. Targeted Audience Engagement

Instead of employing a one-size-fits-all approach, a targeted approach allows you to tailor your marketing efforts to specific audience segments, increasing the relevance of your messaging and the likelihood of conversion. Learn how matching your content marketing efforts to your target audience generates engagement in Chapter 6: "Content Marketing - More Than Just Blogs."

3. Efficient Resource Allocation

By outlining your marketing tactics and allocating resources accordingly, you won't waste time and money on ineffective channels or campaigns. This book recommends numerous tools and tips to ensure marketing efficiencies.

4. Measurable Results

A strategy establishes clear goals and KPIs that enable you to measure the impact of your marketing efforts accurately. It ensures that your marketing investments deliver a tangible ROI. In Part V: Measure and Monitor, we dive into the importance of Data Analytics, Conversion Rate Optimization, Customer Retention and Lifetime Value, and ROI and KPIs.

5. Competitive Edge

In the fiercely competitive field of aviation training, a meticulously executed digital marketing strategy can serve as your flight school's distinctive edge, propelling it to a prominent position within the industry. Amidst the multitude of flight schools vying for the attention of aspiring pilots and aviation enthusiasts, a well-structured digital marketing strategy is akin to the beacon of a runway guiding aircraft safely home.

By crafting and implementing a digital marketing strategy that resonates with your target audience, addresses their unique needs, and showcases the unparalleled value your flight school offers, you not only stand out but also create a lasting impression. This strategy becomes the proverbial launchpad, propelling your flight school to the forefront of the industry's collective consciousness.

Furthermore, an effectively executed digital marketing strategy is a testament to your school's commitment to innovation and adaptability in an ever-evolving industry. It demonstrates your willingness to meet your audience where they are captivating content.

In essence, your digital marketing strategy is the compass that not only sets you apart from the competition but also positions your flight school as a trusted leader, paving the way for sustained growth, enhanced brand recognition, and an exceptional ROI in the dynamic world of aviation training.

YOUR DIGITAL MARKETING STRATEGY **IS THE COMPASS** THAT SETS YOU APART FROM THE COMPETITION.

YOUR BUSINESS PLAN

Your digital marketing strategy should be a supporting pillar that helps you achieve the objectives outlined in your flight school's business plan. Here's how they are interrelated:

1. Goal Alignment

The goals you set in your digital marketing strategy should directly align with the broader goals and objectives included in your business plan. For example, if your business plan aims to increase student enrollments by a certain percentage, your digital marketing strategy should include specific tactics to achieve this.

2. Resource Allocation

Your digital marketing strategy dictates how you allocate resources, whether it's budget, time, or manpower. This allocation should be in harmony with the resource allocation section of your business plan.

3. Audience Targeting

The audience definition and segmentation in your digital marketing strategy should mirror the target audience identified in your business plan. It ensures your marketing efforts are directed towards those most likely to become your students.

4. Timelines and Milestones

The timeline and milestones in your digital marketing strategy should complement the ones set in your business plan. This creates a synchronized approach towards achieving your flight school's objectives within a specified timeframe.

5. Continuous Improvement

Both your business plan and digital marketing strategy should emphasize the importance of regular evaluation and improvement. This iterative process ensures your strategies remain adaptable and responsive to changing market conditions.

STEPS TO CREATE A DIGITAL MARKETING STRATEGY

Now, let's delve into the essential steps to craft a comprehensive digital marketing strategy tailored to your flight school:

1. Define Your Goals

Start by clearly outlining your marketing objectives. Whether it's increasing student enrollments, boosting website traffic, or enhancing brand awareness, your goals should be specific, measurable, achievable, relevant, and time-bound (SMART).

DIGITAL MARKETING IS ADAPTABLE AND RESPONSIVE.

2. Understand Your Audience

Conduct in-depth audience research to create detailed buyer personas. Understand their needs, preferences, pain points, and behaviors. This knowledge forms the foundation of your marketing strategy.

3. Analyze the Competition

Study your competitors to identify gaps in the market, areas where you can differentiate your flight school, and successful strategies you can learn from.

4. Select Your Marketing Channels

Choose the digital marketing channels that are most suitable for reaching your target audience. These may include social media, search engine optimization (SEO), pay-per-click (PPC) advertising, email marketing, content marketing, and more.

5. Create Content Strategy

Develop a content strategy that aligns with your audience's needs and preferences. Consider the balance of messaging that will resonate with them—such as educational, inspirational or promotional content—and the formats they are most likely to consume, which may include a combination of blog posts, videos, webinars, and social media posts.

6. Set a Budget

Allocate a budget for your digital marketing efforts, taking into account the cost of advertising, content creation, tools, subscriptions, and personnel.

7. Develop a Tactical Plan

Outline the specific tactics and activities you will undertake to achieve your goals. Create a content calendar, decide on ad campaigns, and plan your SEO strategy.

8. Implement and Monitor

Execute your digital marketing tactics and closely monitor their performance. Use analytics tools to track key metrics and adjust your approach as needed.

9. Evaluate and Optimize

Regularly assess the results of your digital marketing efforts against your KPIs. Identify what's working and what's not and make data-driven adjustments to optimize your strategy.

YOUR AUDIENCE DEFINES YOUR DIGITAL MARKETING STRATEGY.

INFLUENCE OF AUDIENCE DEFINITION ON YOUR STRATEGY

As we covered in Chapter 2, audience definition plays a pivotal role in shaping every aspect of your digital marketing strategy. Here's how it influences various elements:

1. Content Creation

Knowing your audience's preferences and pain points informs the content you create. It ensures your messaging resonates and addresses their specific needs.

2. Channel Selection

Factors such as audience demographics and behaviors guides your choice of marketing channels. If your audience primarily engages on social media, focusing your efforts there will yield better results.

UNDERSTAND YOUR AUDIENCE, CONTROL YOUR MARKETING RESULTS.

3. Messaging and Tone

Understanding your audience's characteristics helps you tailor your messaging and tone to match their expectations and communication style.

4. Timing and Frequency

Audience habits dictate the timing and frequency of your marketing efforts. Knowing when your audience is most active allows you to schedule posts or ads for maximum impact. Spikes and dips in engagement can signal a need to adjust the frequency of your message delivery.

5. Ad Targeting

Audience definition is crucial for precise ad targeting. It ensures that your advertisements are shown to individuals who are most likely to convert, when they're ready to commit.

REVIEW AND ONGOING OPTIMIZATION

A digital marketing strategy is not set in stone; it's a dynamic framework that requires regular review and optimization. Here's why this step is indispensable:

1. Changing Landscape

The digital marketing landscape is constantly evolving, with new technologies, platforms, and consumer behaviors emerging. Regular reviews help you adapt your strategy to accommodate these changes and stay relevant.

2. Performance Evaluation

Continuously assessing the performance of your digital marketing efforts against your KPIs provides valuable insights. It allows you to identify areas of improvement and to allocate resources effectively.

A WELL-CRAFTED DIGITAL MARKETING STRATEGY IS A CORNERSTONE OF **SUCCESS FOR YOUR FLIGHT SCHOOL.**

3. Competitor Analysis

Your competitors may alter their strategies, creating opportunities or challenges for your school. Staying updated through regular reviews helps you respond strategically to these changes, as well as to new competitors entering the market.

4. Audience Shifts

Audience preferences and behaviors can change over time. Periodic audience research ensures your strategy remains aligned with your audience's current needs and allows you to adjust as those needs change.

5. Experimentation

Regular reviews enable you to experiment with new tactics or channels and test their effectiveness. This spirit of innovation can lead to breakthroughs in your digital marketing strategy.

A well-crafted digital marketing strategy is a cornerstone of success for your flight school. It aligns your marketing efforts with your business plan, guides your actions, and ensures that you remain adaptable in a dynamic digital landscape. By understanding your audience, setting clear goals, and utilizing data-driven approaches, you can not only reach but also exceed your ROI targets. From leveraging social media platforms to diving into the nuanced worlds of SEO, PPC, and email automation, a comprehensive digital strategy allows you to make the most of each interaction with potential and existing students. It's not just about adopting the latest trends; it's about knowing which methods resonate with your specific audience, and then scaling those successful tactics to maximize results.

LEVERAGE YOUR MARKETING SUCCESS

By continuously monitoring performance metrics and being willing to iterate, your flight school can maintain a competitive edge in a crowded marketplace. In essence, a strategic approach to digital marketing is an imperative for sustained growth and profitability.

4

DEFINE, CREATE, AND IMPLEMENT YOUR LONG-TERM DIGITAL MARKETING STRATEGY

CHAPTER FOUR

PLAN FOR THE FUTURE

INTEGRATE AI AND NEW TECH INTO YOUR STRATEGY

In a world where change is the only constant, staying ahead of emerging trends and technologies is a necessity. For aviation training centers and flight schools, digital marketing represents a complex but opportunity-laden space. By keeping an eye on future developments, organizations can position themselves at the forefront of the industry. A study by Gartner highlights that early adopters of new technologies can gain an edge, achieving as much as a 25% increase in ROI. This chapter explores the upcoming trends and technologies in digital marketing that could significantly impact the aviation training sector.

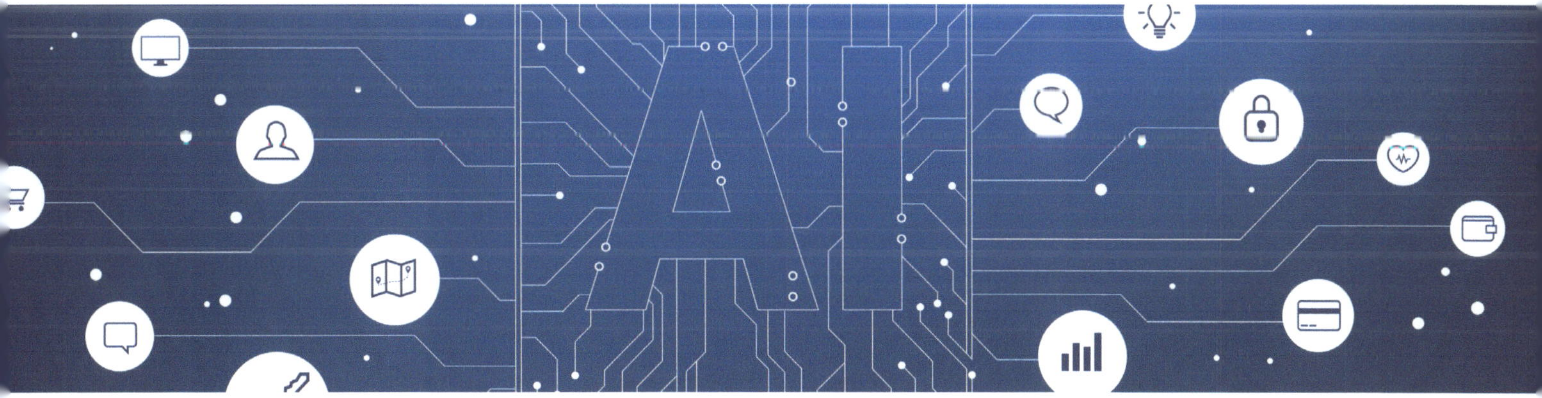

THE RISE OF ARTIFICAL INTELLIGENCE

TRANSFORMING CUSTOMER INTERACTIONS

As we step into the future of business, it's evident that Artificial Intelligence (AI) is on the cusp of redefining the way we engage with our customers. One of the most striking manifestations of this transformation is the rise of chatbots. These AI-driven virtual assistants have rapidly evolved from being a mere novelty to becoming essential tools for businesses across various industries. In this chapter, we'll examine how AI—particularly chatbot—is poised to revolutionize customer interaction, enhancing efficiency and customer satisfaction.

CHATBOTS ARE NO LONGER A **FUTURISTIC CONCEPT**, BUT A PRESENT-DAY REALITY

Chatbots, the digital customer service agents powered by AI, are no longer a futuristic concept but a present-day reality. They can handle a wide range of tasks, from addressing basic queries to facilitating complex transactions. Their ability to swiftly and accurately respond to customer inquiries, 24/7, is unparalleled. This round-the-clock availability ensures businesses can cater to their customers' needs at any hour, enhancing the overall customer experience.

The adoption of AI in marketing is well underway. Salesforce states that 51% of marketers are already leveraging AI in their operations, and an additional 27% are actively planning to incorporate this transformative technology by the end of next year. This statistic alone underscores the growing importance of AI in reshaping the landscape of customer interaction.

THE POWER OF AI-DRIVEN CHATBOTS

One of the most compelling aspects of AI-driven chatbots is their versatility. These digital assistants can perform a multitude of tasks, each contributing to more seamless and efficient customer interaction. Let's explore some essential functions chatbots excel at:

1. Instant Responses to Customer Queries

Gone are the days of customers waiting impatiently for a response. Chatbots provide immediate answers to frequently asked questions, ensuring customers get the information they need without delay. This enhances customer satisfaction and frees up human agents to focus on more complex tasks.

2. Appointment Scheduling and Reminders

Chatbots are adept at handling appointment bookings and sending reminders. Whether setting up a doctor's appointment, scheduling a service call, or booking a table at a restaurant, chatbots can streamline the process, reducing the chances of errors and no-shows.

3. Personalized Recommendations

AI-driven chatbots are excellent at analyzing customer data and preferences. They can offer customized product recommendations—a level of personalization that makes customers feel valued and understood, and enhances cross-selling and upselling opportunities.

4. Handling Transactions

Chatbots can facilitate financial transactions, making it easy for customers to make purchases or payments within the chat interface. This is particularly valuable in e commerce, where seamless transactions can significantly boost sales.

5. 24/7 Support

The beauty of chatbots is that they never sleep. They are available around the clock to assist customers, ensuring that no query or concern goes unanswered, regardless of the time of day or night. This accessibility strengthens customer satisfaction, loyalty and trust.

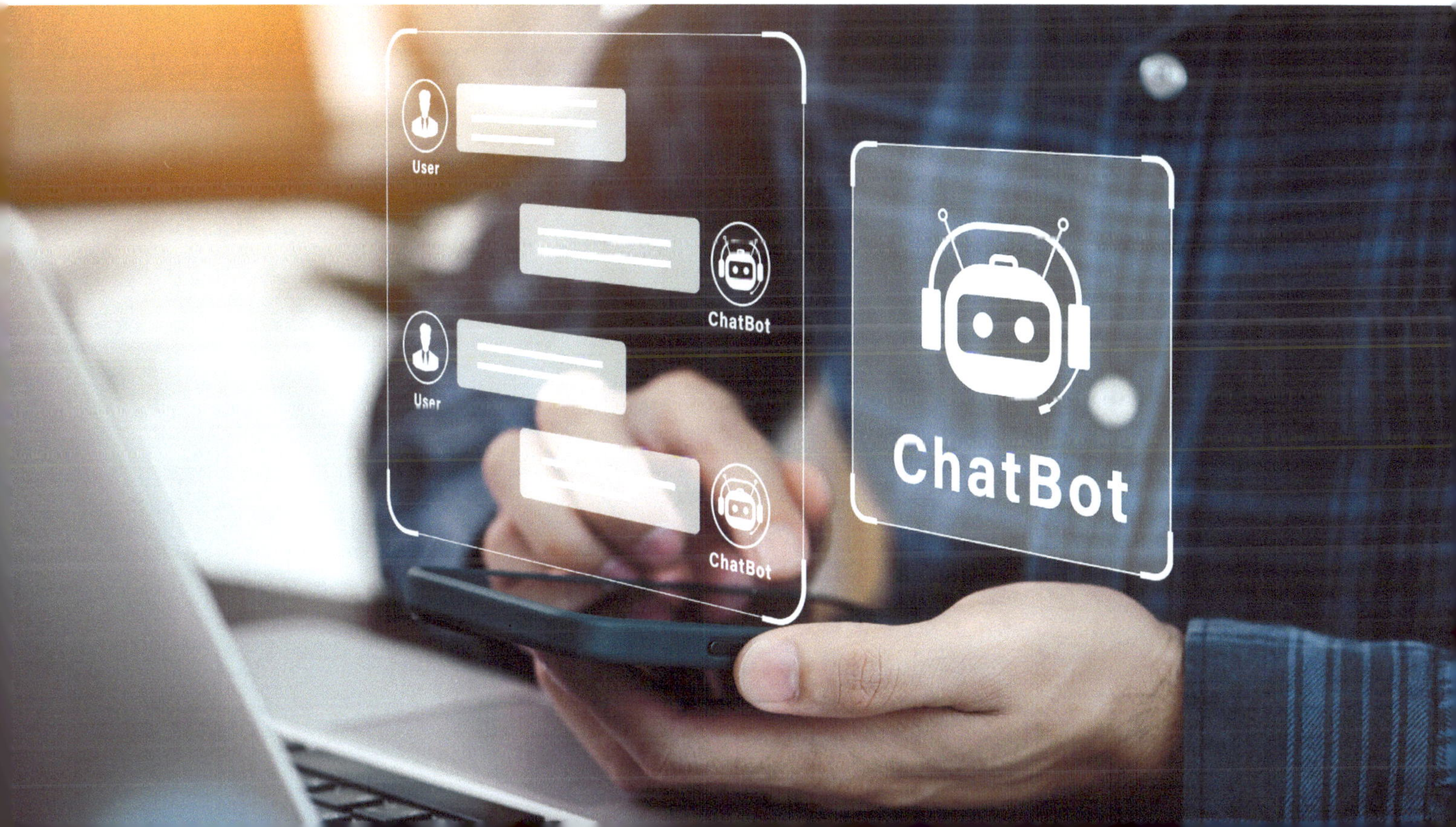

THE DATA-DRIVEN ADVANTAGE

Another compelling aspect of AI-driven customer interactions is the wealth of data they generate. Every interaction with a chatbot provides valuable customer insights. This data can be harnessed to refine marketing strategies, improve products and services, and tailor future interactions to be even more customer centric.

For instance, businesses can identify common customer issues and pain points by analyzing chatbot interactions. This information can be used to proactively address these concerns, preventing potential problems and enhancing customer satisfaction.

AI can also segment customers based on their behavior and preferences, allowing businesses to deliver highly targeted messaging. This level of personalization not only improves the effectiveness of marketing efforts but also strengthens the bond between customers and brands.

THE ROAD AHEAD: HARNESSING AI FOR COMPETITIVE ADVANTAGE

As we look to the future, it's clear that AI-driven customer interaction is not just a passing trend but a transformative force that will shape the business landscape for years to come. To harness the full potential of AI, businesses need to take a proactive approach:

1. Embrace AI Integration

Businesses should begin integrating AI into their customer interaction strategies. Whether deploying chatbots, automating marketing campaigns, or utilizing AI for data analysis, the key is to leverage AI to enhance efficiency and customer satisfaction.

2. Invest in Training

As AI becomes more integrated into customer interactions, it's essential to ensure your employees are trained to work alongside these technologies effectively. Human-AI collaboration will be a critical skillset in the workforce of the future.

3. Continual Adaptation

AI is a dynamic field, and technologies evolve rapidly. Businesses must stay updated with the latest AI advancements to remain competitive. Regularly evaluating and upgrading AI solutions is crucial.

The adoption of AI in customer interaction is well underway, and it's a journey no business can afford to miss. With their ability to provide instant responses, streamline processes, and generate valuable data, chatbots are at the forefront of this revolution. By embracing AI, businesses can enhance customer satisfaction and gain a competitive edge in an increasingly digital world. The future of customer interaction is here, driven by the power of Artificial Intelligence.

PREDICTIVE ANALYTICS

In its ever-evolving role within businesses, AI has transcended the simple task of automating repetitive processes. While automation remains a valuable aspect of AI, its true power lies in its capacity for intelligent data analysis. This transformative capability is made possible by applying machine learning algorithms, which enable AI to do much more than follow predetermined instructions. It empowers AI to unearth patterns, and predict future consumer behavior, revolutionizing the way companies approach marketing.

At the heart of this AI-driven transformation is the concept of predictive analytics. Rather than relying solely on historical data to make decisions, businesses can now leverage AI to forecast future trends and consumer actions. This is invaluable in marketing, where understanding and influencing consumer behavior is paramount.

> WHEN CONSUMERS RECEIVE CONTENT AND OFFERS **THAT RESONATE WITH THEIR NEEDS AND INTERESTS**, THEY ARE MORE LIKELY TO RESPOND POSITIVELY.

Machine learning algorithms, a subset of AI, excel at recognizing patterns within vast datasets. They can identify subtle correlations and dependencies that might escape human observation. By analyzing historical customer data, AI can examine what consumers have done in the past to predict what they're likely to do in the future. This opens up a world of possibilities for targeted marketing strategies.

One of the key benefits of AI-driven predictive analytics in marketing is the ability to personalize and optimize campaigns. Businesses can create highly tailored marketing messages and offers based on individual customer preferences, behaviors, and predicted actions. This level of personalization enhances customer engagement and increases the chances of conversion. Consumers are more likely to respond positively when they receive content and offers that resonate with their needs and interests.

AI can continuously adapt and refine its predictions as new data becomes available. This adaptability ensures marketing strategies remain relevant and effective in an ever-changing landscape. For instance, if AI detects a shift in consumer preferences or a new trend emerging, it can automatically adjust marketing campaigns to capitalize on these insights.

Additionally, AI can optimize resource allocation. Businesses can allocate their marketing budget more efficiently by predicting which channels and strategies are likely to yield the best results for specific customer segments. This not only saves resources but also maximizes the return on investment.

AI IS AN **ESSENTIAL TOOL** FOR COMPANIES LOOKING TO THRIVE IN A DIGITAL AGE.

AI's role in business extends far beyond automating routine tasks. Its true power lies in intelligent data analysis and predictive capabilities. Through machine learning algorithms, AI can analyze vast datasets, identify patterns, and forecast consumer behavior. This is a game-changer in marketing, enabling businesses to create highly targeted and personalized campaigns, optimize resource allocation, and stay agile in a dynamic market. As AI continues to evolve, its impact on marketing and other areas of business will only become more profound, making it an essential tool for companies looking to thrive in the digital age.

VIRTUAL REALITY (VR) AND AUGMENTED REALITY (AR)

Real-World Simulations

The aviation industry's pioneering spirit extends not only to its groundbreaking technologies in the skies but also to its innovative applications of Virtual Reality (VR) and Augmented Reality (AR) in training and marketing. While VR and AR have played a pivotal role in aviation training for years, their utilization is now expanding into marketing, promising to transform how companies engage with their audiences.

THE INTEGRATION OF VR AND AR INTO AVIATION MARKETING **OPENS UP OPPORTUNITES** FOR INTERACTIVE MARKETING CAMPAINGS.

VR and AR have been instrumental in simulating real-world flight scenarios, providing trainees with immersive experiences that enhance their skills and situational awareness. Pilots can practice emergency procedures in a safe and controlled virtual environment, and maintenance technicians can learn to troubleshoot complex systems using augmented reality overlays. This technology improves training efficiency and contributes to enhanced safety standards within the aviation industry.

However, the influence of VR and AR isn't confined solely to the cockpit. These technologies are now entering the marketing arena, where they are poised to revolutionize customer engagement. Imagine a prospective traveler using an AR app to visualize their in-flight experience, exploring the aircraft cabin layout, seat options, and in-flight entertainment offerings. This immersive preview not only excites potential customers but also helps airlines market their services more effectively.

Goldman Sachs forecasts, the VR and AR market in education, which encompasses aviation training, will burgeon to a remarkable $700

million by 2025. This growth reflects the growing recognition of the potential for VR and AR to enhance learning experiences and, by extension, marketing efforts. As aviation training and marketing increasingly converge, the industry stands to benefit from innovative applications that span from pilot instruction to passenger engagement.

The integration of VR and AR into aviation marketing not only promises to enrich customer experiences and create opportunities for interactive advertising campaigns. Airlines and aviation companies can create immersive brand experiences that allow customers to virtually explore destinations, cabin configurations, and onboard services, providing a more engaging and informative decision making process.

The aviation industry's adoption of VR and AR is extending beyond training to embrace marketing. As the VR and AR market in education continues its ascent, aviation companies are well-positioned to leverage these technologies to create more compelling marketing strategies. The future of aviation marketing is undoubtedly set to be shaped by the immersive experiences afforded by VR and AR, enhancing customer engagement and driving the industry forward.

Enhancing Campus Tours

Imagine giving prospective students a virtual tour of your training facilities, complete with a VR flight simulation. This immersive approach to marketing can significantly boost enrollment rates by offering a firsthand glimpse into the exciting world of aviation training.

Through VR, aspiring pilots can step into your state-of-the-art classrooms, see the meticulously maintained aircraft, and even sit in the cockpit for a virtual flight lesson. They can experience the thrill of takeoff and landing while gaining a sense of the professionalism and expertise your aviation training program offers. This virtual tour not only informs but also captivates potential students, igniting their passion for aviation and increasing the likelihood of them choosing your institution for their training.

VR tours transcend geographical boundaries, allowing prospective students worldwide to explore your facilities without requiring physical travel. This accessibility broadens your reach and attracts a diverse pool of candidates, enriching the training community and fostering a global perspective among your students.

The impact of such immersive marketing extends beyond enrollment numbers. It creates a lasting impression of your training program, setting the stage for students to embark on a journey filled with excitement, challenge, and accomplishment. As they transition from virtual exploration to hands-on training, the initial sense of wonder cultivated through VR can heighten their commitment to their aviation education, leading to more dedicated and successful students.

Embracing VR for marketing in aviation training offers a powerful tool to engage and inspire prospective students. Providing immersive virtual experiences, helps you attract a broader audience and lay the foundation for a future generation of passionate and dedicated aviation professionals. It's a compelling way to showcase the excellence and opportunities your training program offers, ultimately elevating your institution's reputation and impact within the aviation community.

VOICE SEARCH OPTIMIZATION

Speak to Search

As smart speakers like Amazon Alexa and Google Assistant become more popular, optimizing for voice search is critical. Market research and analysis firm Market.us predicts the global voice search market will grow from an estimated $20.3 billion in 2022 to $112.5 billion by 2032.[17]

Keywords and Natural Language

Voice search often involves natural language queries. This requires a shift in your SEO strategy, specifically choosing keywords, focusing on conversational phrases rather than traditional keywords.

CONCLUSION

New technologies and trends are redefining the boundaries of what's possible in digital marketing for aviation training centers. By investing in these emerging trends, flight schools can improve their ROI while offering an enhanced experience to their students. According to a survey by Deloitte, companies at the forefront of technological adoption are 3.1 times more likely to outperform their peers in terms of profitability.

AI MARKETING CHECKLIST

- ☐ Investigate AI solutions suitable for customer interaction and data analysis.
- ☐ Explore the feasibility of incorporating VR and AR into your marketing strategies.
- ☐ Assess the applicability of blockchain technology in your business operations.
- ☐ Revisit your SEO strategy to include voice search optimization.

PART II

TACTICS

5

UNDERSTAND WHAT MAKES A SUCCESSFUL WEBSITE AND LEARN HOW TO PLAN YOUR OWN

CHAPTER FIVE

Successful Website Design

ANATOMY OF A HIGH-CONVERTING WEBSITE FOR FLIGHT AND GROUND SCHOOLS

Your website is often the first point of contact between your flight or ground school and potential students. Thus, making this digital first impression as impactful as possible is critical. CRO is gaining prominence in digital marketing parlance. In essence, CRO is the science and art of increasing the percentage of visitors to a website who take desired actions—such as submitting an inquiry, signing up for a newsletter, or enrolling in a course. An Adobe study confirmed that companies with a well-structured testing program (a cornerstone of CRO) enjoy a 60% higher conversion rate. Therefore, it's not just about attracting eyeballs; it's about converting those eyeballs into action-takers and, ultimately, students or clients.

USER EXPERIENCE (UX)

The first element to consider is the user experience. A clean design, easy navigation, and clear calls to action (CTAs) can significantly improve conversions. In fact, a study by Forrester found that a well-designed user interface could yield conversion rate improvements of up to 200%. Seemingly minor details like using contrasting colors for your CTA buttons and strategically placing them—typically above the fold for immediate visibility and at the end of important content—can yield substantial gains in conversions.

DO'S FOR GOOD UX

- **Prioritize UX:** Make user experience a central focus of your design and development process.
- **Clean Design:** Opt for a clean and clutter-free design that enhances navigation and visual appeal.
- **Enhance Readability:** Keep copy concise, using shorter paragraphs and bulleted lists where possible.
- **Easy Navigation:** Ensure intuitive navigation with clearly labeled menus and easy-to-understand pathways through the website.
- **Clear CTAs:** Create CTAs that guide users toward desired actions.
- **Utilize Contrasting Colors:** Use contrasting colors for CTA buttons to make them stand out and grab users' attention.
- **Strategic CTA Placement:** Position CTAs strategically, often above the fold for immediate visibility and at the end of important content to capitalize on user engagement.
- **Optimize for Mobile:** Ensure readability on a smaller screen and make icons and links large and far enough apart for easy one-handed clicking.

DON'TS FOR GOOD UX

- **Neglect UX:** Don't sideline UX considerations; it should be at the forefront of your design strategy.
- **Cluttered Design:** Avoid cluttered and overwhelming designs that confuse or distract users.
- **Complex Navigation:** Steer clear of complex navigation structures that confuse users or make it difficult to find what they're looking for.
- **Ambiguous CTAs:** Don't use vague or uninspiring CTAs; they should clearly convey the action users should take.
- **Poor Color Choices:** Avoid using colors that don't contrast well or clash, as this can make CTAs less noticeable.
- **Haphazard Placement:** Don't randomly place CTAs; instead, strategically position them to guide users through the conversion journey.

By following these dos and don'ts, you can create a user-friendly website that not only enhances the user experience but also improves conversion rates and overall satisfaction.

EXAMPLES OF GOOD UX

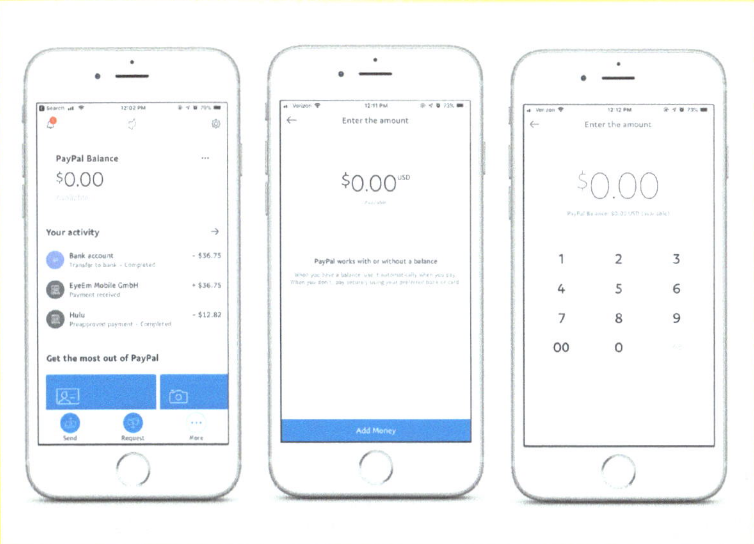

PAYPAL

Paypal's simple, clean, and easy to navigate layout makes this a very successful UX design. Function is incredibly important for user retention and satisfaction. In this case, Paypal combines excellent functionality with solid design.

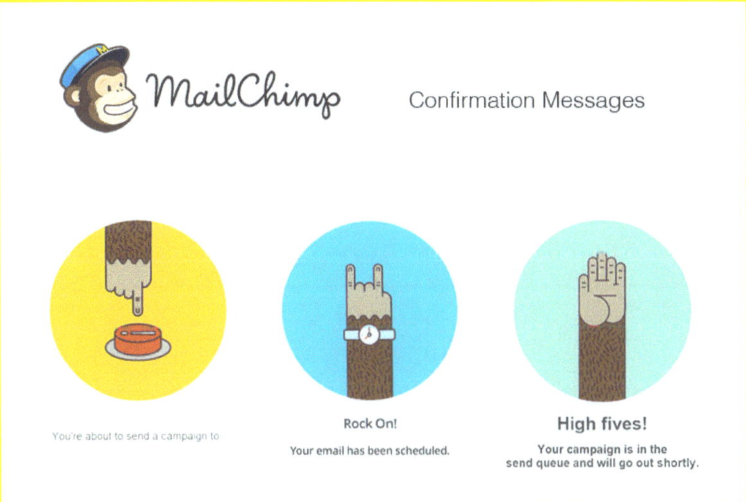

MAILCHIMP

MailChimp's use of their mascot, Frederick von Chimpenheimer, personalizes their site. MailChimp then utilizes all opportunities to highlight this personal connection, for example, by using the confirmation messages as a vehicle for expression.

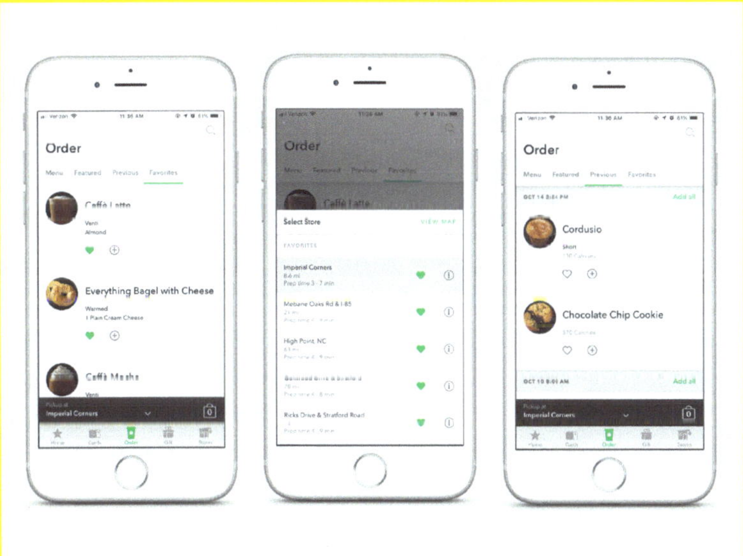

STARBUCKS

Starbucks utilizes AI in their UX, which is another way to personalize the experience. The site uses AI to understand the purchase history and patterns of their customers, then customizes the "Featured" tab for each user.

Images: https://trone.com/blog/10-examples-good-user-experience-ux

Content Strategy

High-quality, targeted content can educate, inform, and convert. Creating persona-specific landing pages that address the unique needs of your different audience segments can ensure you direct each user to the content that's most relevant to them. For instance, for professionals seeking advanced training, you could have landing pages discussing your state-of-the-art simulators and FAA-compliant programs. We'll dive into this in more detail in Chapter 6: "Content Marketing."

THE POWER OF TRUST INDICATORS IN AVIATION TRAINING MARKETING

DISPLAYING REVIEWS CAN INCREASE CONVERSION BY **270%**

Testimonials, reviews, and trust badges can go a long way in establishing credibility. According to Spiegel Research Center, displaying reviews can increase conversion by 270%. Consider showcasing student testimonials, FAA certifications, and any awards you might have received.

In the digital age, where choices abound, trust is a currency that can make or break a business. Aviation training companies, in particular, thrive on trust. Prospective student pilots must have confidence in the quality of training, the safety of the instructors, and the reputation of the institution. In this context, testimonials, reviews, and trust badges are pivotal in establishing credibility and attracting new students. The impact of these trust indicators is not to be underestimated, as they can increase conversion rates significantly.

The Trust-Building Power of Testimonials

Testimonials are a powerful way to showcase real-life experiences of students who have undergone training with your institution. These personal accounts provide a human touch to your marketing efforts, making your aviation training program relatable and trustworthy.

Research conducted by BrightLocal, a marketing insights company, reveals that 91% of consumers aged 18-34 trust online reviews as much as personal recommendations. This statistic underscores the influence of testimonials in shaping consumer decisions. When prospective student pilots read about the successes and positive experiences of your past students, they are more likely to trust your institution with their aviation dreams.

Leveraging the Impact of Reviews

Online reviews have become a cornerstone of decision-making in the digital age. Whether it's selecting a restaurant, booking a hotel, or choosing an aviation training program, people turn to reviews for insights and recommendations.

The Spiegel Research Center's finding that displaying reviews can increase conversion rates by a staggering 270% emphasizes the incredible potential of reviews in driving business growth. In the context of aviation training, this means that showcasing reviews from satisfied students can be a game-changer.

However, the importance of reviews goes beyond mere quantity; quality matters too. Encourage students to provide detailed and authentic reviews that highlight specific aspects of their learning journey, the quality of instructors, and the positive outcomes they've achieved. These in-depth reviews carry more weight and provide valuable insights for potential students.

Trust Badges: Symbols of Authority and Security

Trust badges, such as FAA certifications, awards, and affiliations, are symbols of authority and security. These badges reassure potential students your institution meets industry standards and operates with integrity.

Consider this scenario: A prospective student is browsing aviation training options online. They visit your website, and see an FAA certification badge prominently displayed on the homepage. Instantly, they feel a sense of trust and credibility, knowing that your institution complies with stringent aviation regulations.

A study by the Baymard Institute, a web usability research organization, found that trust badges addressing security concerns, such as SSL certificates, can increase conversion rates by up to 42%. While this specific statistic pertains to e-commerce, the principle remains relevant – trust badges instill confidence.

The Psychological Impact of Trust Indicators

To understand the significance of trust indicators, it's essential to examine the psychology behind them. Trust is a fundamental aspect of human decision-making. When faced with choices, individuals often rely on mental shortcuts or heuristics to simplify the decision-making process. Trust indicators are an example of these shortcuts.

For instance, when a potential student visits your website and sees a collection of glowing testimonials from past students, their brain registers this as social proof. It signals that others have already decided to trust your institution, making it a safer choice. This phenomenon is known as the bandwagon effect, where people tend to follow the actions of the crowd.

Similarly, reviews tap into the principle of consensus. When numerous students share positive experiences, it reinforces the idea that your aviation training program is a trusted and popular choice.

Trust badges appeal to the need for authority and credibility. They evoke the concept of expertise and competence, assuring potential students your institution possesses the necessary qualifications and certifications.

The Role of Trust in the Aviation Training Sector

In the aviation training sector, trust is non-negotiable. Potential student pilots are making a significant investment, of their time, commitment and money. They want assurance that their training will be of the highest quality, that the instructors are skilled and experienced, and that the institution has a proven track record of success.

Testimonials, reviews, and trust badges serve as beacons of trust in this industry. They help answer crucial questions in the minds of prospective students:

1. Is this institution reputable?

Testimonials from successful graduates and positive reviews provide evidence of your institution's reputation.

2. Can I trust the instructors?

Reviews that mention the expertise and supportiveness of your instructors reassure potential students.

3. Is this institution compliant with aviation regulations?

FAA certifications and other trust badges demonstrate your commitment to safety and industry standards.

4. Will I succeed here?

When past students share their achievements and growth, it inspires confidence in the potential for success within your program.

Testimonials, reviews, and trust badges wield the power to build trust, inspire confidence, and ultimately drive conversions. They are not mere marketing tools but essential elements that bridge the gap between prospective students' aspirations and the realization of their aviation dreams. By leveraging these trust indicators effectively, aviation training institutions can create a compelling narrative of credibility and excellence, attracting students who are not just seeking an education but a trustworthy partner in their aviation journey.

SPEED AND MOBILE RESPONSIVENESS

With Google's mobile-first indexing, your site must be optimized for mobile devices. A one-second delay in page response can result in a 7% reduction in conversions, says data from Akamai. Make sure your site loads quickly on all devices to provide a seamless user experience.

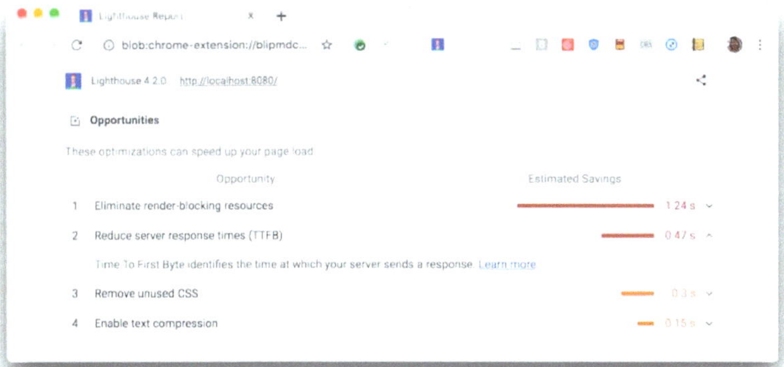

Speed Index (in seconds)	Color-coding
0–3.4	Green (fast)
3.4–5.8	Orange (moderate)
Over 5.8	Red (slow)

LIGHTHOUSE - Website Optimization Tool

Lighthouse is an open-source to that allows you to not only optimize the speed of your website but also to control other aspects of your site to enhance and optimize your traffic.

- Performance Audits
- Accessibility Audits
- Best Practices Audits
- SEO Audits
- PWA Audits

This free tool gives you the ability to control your website's performance which directly affects your marketing success.

LEAD CAPTURE MECHANISMS

These can be as straightforward as newsletter sign-ups or as involved as a multi-step quiz that recommends courses based on the user's responses. Both of these are effective ways to get users to opt-in to receive marketing emails from your school. A study by VentureBeat states that, email marketing has an ROI of $38 for every $1 spent, making your email list an invaluable asset.

EXAMPLES OF LEAD CAPTURE MECHANISMS

ENTREPRENEURS ON FIRE

Mailing List Sign Up

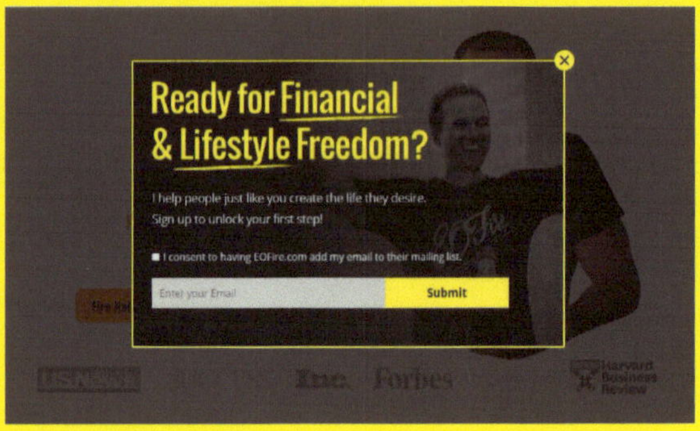

MIKELARDI.COM

Free Download

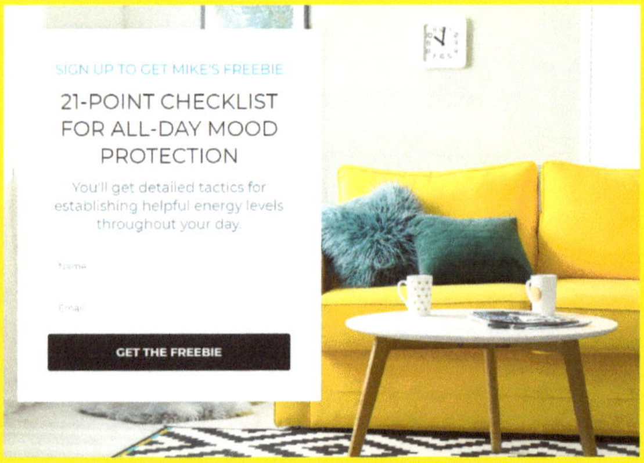

YUMBLE

Customize Search Results

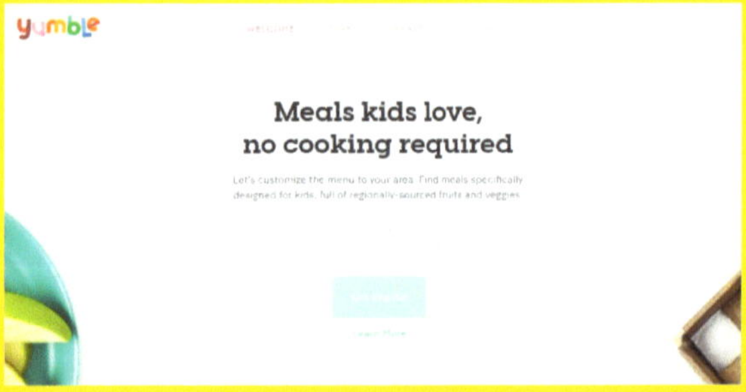

CUP & LEAF

Product Quiz

Images:
https://wpforms.com/lead-generation-forms-examples/
https://www.drip.com/blog/quiz-funnel-examples

ANALYTICS AND TESTING

Optimizing Conversion Rates Through Data-Driven Strategies

In the ever-evolving landscape of digital marketing, understanding user behavior and making data-informed decisions is paramount. Continuously gathering data using tools like Google Analytics (GA) provides invaluable insights into how users interact with your website. One critical aspect of this data-driven approach is identifying points of user drop-off and implementing strategies to enhance conversion rates.

The Power of Google Analytics

GA is a treasure trove of data that enables you to gain a comprehensive understanding of your website's performance. It tracks user behavior, traffic sources, and engagement metrics, helping you pinpoint areas that require improvement. You can address pain points and streamline the conversion process by analyzing user flow and identifying where visitors drop off or abandon their journey.

For instance, if you notice a significant drop-off on a particular page, GA can help you uncover the reasons behind this phenomenon. It could be slow loading times, unclear navigation, or a lack of engaging content. Armed with this information, you can make informed adjustments to improve user experience and boost conversions.

A/B Testing for Precision Optimization

A/B testing, also known as split testing, is a dynamic strategy that allows you to experiment with variations of elements on your website to determine which ones yield the best results. Features commonly tested include headlines, CTA buttons, forms, images, and even entire page layouts.

For instance, you might run an A/B test to assess the impact of two headline variations on your landing page. By randomly showing version A or version B to users, you can analyze which headline generates higher engagement and conversion. This data-driven approach enables you to refine your content and design based on what resonates most with your audience.

The Impact of A/B Testing on Conversion Rates

A/B testing is not just a hypothetical strategy; real-world results support its effectiveness. A study from Visual Website Optimizer found that businesses implementing A/B testing often report customer conversion rates 2x higher than the industry average. This remarkable statistic underscores the significance of A/B testing as a powerful tool for optimizing websites and marketing campaigns.

A/B testing is an ongoing process. It's not a one-time endeavor but a continuous cycle of refinement. As you gather more data and insights, you can make incremental improvements to your website and marketing strategies, gradually enhancing conversion rates over time.

CONCLUSION

The Data-Driven Path to Conversion Rate Optimization

In a digital landscape where user attention spans are fleeting, and competition is fierce, data-driven strategies are crucial for success. Continuously gathering data through tools like Google Analytics allows you to identify drop-off points and bottlenecks in the user journey. A/B testing, on the other hand, empowers you to refine elements like headlines, CTAs, and forms to improve conversion rates.

As evidenced by the VWO study, A/B testing is a proven method for businesses to achieve customer conversion rates well above the industry average. This highlights the transformative potential of data-driven optimization in driving business growth and improving the overall user experience.

The path to higher conversion rates is paved with data-driven decisions and A/B testing. By leveraging these strategies, flight schools can enhance their digital presence and increase their ability to convert visitors into satisfied students, ultimately leading to greater success in the competitive online landscape.

HIGH-CONVERTING WEBSITE CHECKLIST

- ☐ **Clear, Actionable CTAs:** Make sure your CTAs stand out and tell the visitor exactly what action they should take next.
- ☐ **Personalized Landing Pages:** Create different landing pages for different audience segments.
- ☐ **Trust Indicators:** Showcase reviews, testimonials, and certifications.
- ☐ **Speed Optimization:** Test your website's speed and improve it for both desktop and mobile devices.
- ☐ **Mobile Responsiveness:** Ensure your website is easily navigable on mobile devices.
- ☐ **Lead Capture Mechanisms:** Integrate sign-up forms, quizzes, or other lead capture tools into your site.
- ☐ **Ongoing Analytics and Testing:** Regularly review analytics data and run A/B tests to optimize performance.

UTILIZE WEBSITE SUCCESS STRATEGIES

By paying attention to these elements, you can successfully convert your website traffic into tangible leads and revenue. It's this tactical approach that will set your flight or ground school apart in a crowded marketplace.

6

UNDERSTANDING SEO IS INTEGRAL TO YOUR MARKETING SUCCESS.

CHAPTER SIX

SEARCH ENGINE OPTIMIZATION

CLIMBING THE RANKS

The digital landscape is flooded with competitors, and your potential students are likely to start their research with a search engine query. Ensuring they can find your flight or ground school online is a necessity. As per a BrightEdge study, over 53% of trackable website traffic comes from organic search, making Search Engine Optimization (SEO) a vital component of your marketing strategy. SEO involves optimizing various elements of your website and online content to appear more prominently in search engine rankings. The goal is to increase visibility and attract quality traffic, which can be converted into students. Successful SEO is a complex blend of technical, on-page, and off-page strategies, often guided by the evolving algorithms of search engines like Google.

UNLOCKING KEY WORD RESEARCH WITH GOOGLE ANALYTICS

A COMPREHENSIVE GUIDE

Before developing a Search Engine Optimization strategy, you need to understand what potential students are searching for. Google's Keyword Planner or tools like SEMrush can offer insights into the search terms people use when researching aviation training. Optimizing your website for relevant keywords will make it easier for the right audience to find you. According to Moz, long-tail keywords account for 70% of all web searches, making them incredibly important for niche businesses like flight schools.

Keyword research is the cornerstone of any effective SEO strategy. It's the process of identifying the specific words and phrases potential students are using when searching for aviation training services. Google Analytics can provide invaluable insights into these keywords, helping you optimize your website to attract the right audience. In this comprehensive guide, we'll walk you through the steps of using GA for keyword research.

Step 1: Set Up Google Analytics

Before you can start gathering keyword data, you need to have GA set up for your website. If you haven't done this yet, follow Google's instructions to create an account, add your website, and install the tracking code. Once it's up and running, you'll be ready to begin your keyword research.

Step 2: Navigate to Google Analytics**

Log in to your Google Analytics account and select the website you want to analyze. From the left-hand menu, go to "Acquisition" and then "All Traffic." Here, you'll find a submenu that includes "Channels," "Source/Medium," and "Referrals."

Step 3: Explore Organic Search Traffic

Click on "Channels" to access the "Organic Search" channel. This section provides data about the keywords that are driving organic traffic to your website.

Step 4: Analyze Keyword Data

In the "Organic Search" section, you'll see a list of keywords along with metrics like sessions, bounce rate, and conversion rate. This data reveals which keywords are bringing visitors to your site and how those visitors are interacting with your content.

Step 5: Filter and Sort Keywords

To refine your keyword research, you can use filters and sorting options within Google Analytics. For example, you can filter keywords by specific landing pages, allowing you to see which keywords are driving traffic to particular sections of your site. Additionally, you can sort keywords by sessions or conversion rate to identify which ones are most impactful.

Step 6: Identify High-Performing Keywords

Look for keywords that drive significant traffic, have a low bounce rate, and contribute to conversions. These are the keywords you want to prioritize in your SEO efforts.

Step 7: Expand Your Keyword List

GA provides data on the keywords that are already driving traffic to your site. However, it's also crucial to identify additional relevant keywords you might not be targeting yet. To do this, you can use external keyword research tools like Google's Keyword Planner.

Step 8: Using Google's Keyword Planner

Google's Keyword Planner is a complementary tool that can help you expand your keyword list. Access it through your Google Ads account by selecting "Tools & Settings" and then "Keyword Planner." Here, you can enter keywords related to aviation training and receive suggestions for additional keywords, along with data on search volume and competition.

Step 9: Incorporate Long-Tail Keywords

Incorporating long-tail keywords is particularly important for niche businesses like flight schools. These are longer and more specific keyword phrases that potential students might use when searching for specific aviation training programs or services. GA and the Keyword Planner can help you identify and prioritize these valuable long-tail keywords.

Step 10: Implement SEO Strategies

With your keyword research in hand, you can now implement SEO strategies to optimize your website. This includes incorporating your chosen keywords into your website's content, meta tags, headers, and even image alt text. Create high-quality, informative content that aligns with the keywords you're targeting, providing value to your audience while improving your search engine rankings.

In addition to Google Analytics, several other online tools offer diverse features and insights for keyword research. SEMrush provides a comprehensive keyword research platform and offers competitive analysis, backlink tracking, and content ideas. Ahrefs excels in backlink analysis and keyword research, aiding in identifying the most relevant keywords for your aviation training business. For those looking for a user-friendly interface, Moz's Keyword Explorer simplifies the process of discovering new keywords and assessing their difficulty. Ubersuggest is a free tool that can generate keyword ideas, while AnswerThePublic specializes in uncovering questions and phrases that users are searching for online. These versatile tools complement each other, allowing aviation training companies to refine their keyword strategies and maximize their online visibility effectively.

ON-PAGE OPTIMIZATION

Elements like meta titles, meta descriptions, and header tags (H1, H2, H3, etc.) should be carefully crafted to include your target keywords. Neil Patel states that a well-optimized meta description can increase click-through rates by up to 5.8%. Make sure that your content is original, relevant, and includes your chosen keywords naturally.

ON-PAGE OPTIMIZATION CHECKLIST FOR IMPROVED SEO

1. Original and High-Quality Content

- ☐ Ensure that your content is unique, informative, and well-researched.
- ☐ Avoid duplicate content, as it can negatively impact your SEO efforts.
- ☐ Regularly update and refresh content to keep it relevant and valuable.

2. Keyword Research

- ☐ Conduct thorough keyword research to identify relevant and high-impact keywords for your aviation training content.
- ☐ Consider long-tail keywords, which are specific and can attract more targeted traffic.

3. Keyword Placement

- ☐ Utilize keywords strategically in the following elements:
    - **Meta Titles:** Craft unique and keyword-rich meta titles for each page, making them compelling and under 60 characters.
    - **Meta Descriptions:** Create engaging meta descriptions that include keywords and encourage click-throughs (under 155-160 characters).
    - **Header Tags (H1, H2, H3, etc.):** Structure your content with clear headers, incorporating keywords in the H1 and H2 tags as naturally as possible.

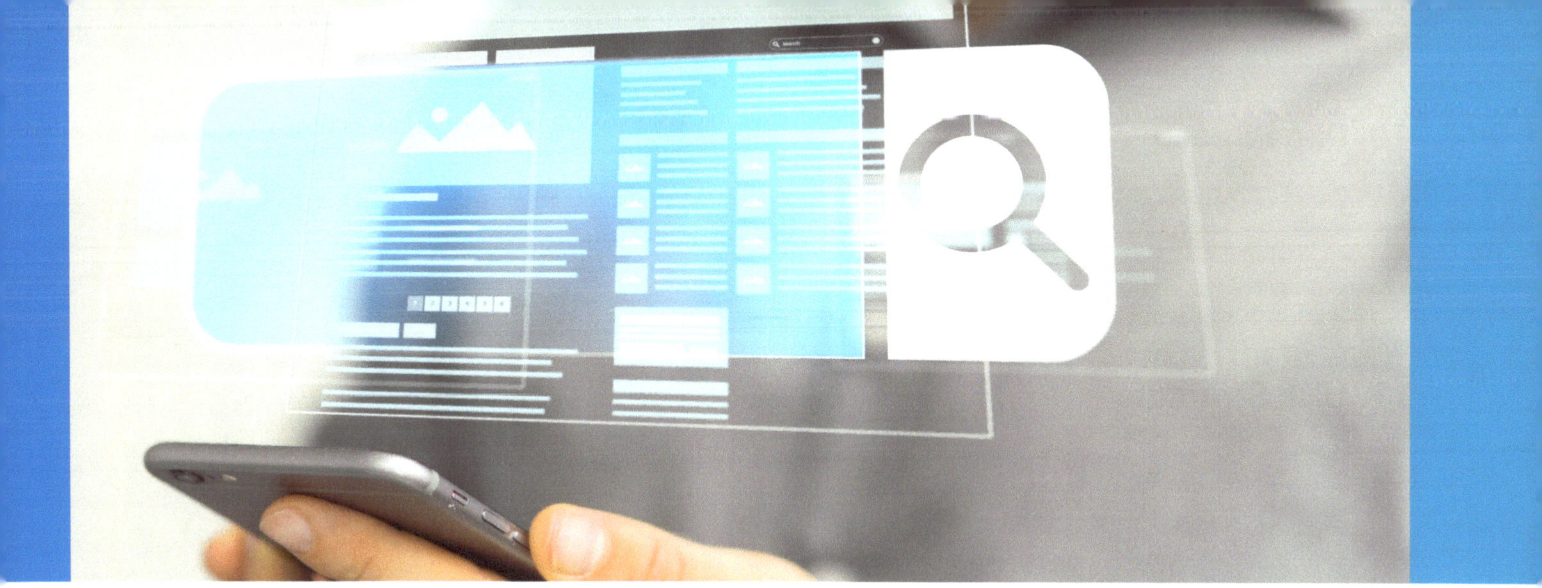

4. Keyword Density

- ☐ Maintain a reasonable keyword density (typically 1-2%) to avoid over-optimization, which can harm your rankings.
- ☐ Use keywords naturally within the content, ensuring readability and user engagement.

5. Regular Updates

- ☐ Continuously monitor and update your on-page optimization efforts based on performance data and changing search trends.

6. Optimize Images

- ☐ Compress and optimize images to improve page load times.
- ☐ Use descriptive alt text for images, including keywords where appropriate.

7. Mobile Optimization

- ☐ Ensure your website is mobile-responsive for a seamless user experience on all devices.
- ☐ Test your site's mobile-friendliness using tools like Google's Mobile-Friendly Test.

8. User Experience (UX)

- ☐ Prioritize user-friendly navigation and a clean, intuitive design.
- ☐ Improve user experience to reduce bounce rates and encourage longer on-page time.

9. Meta Description Optimization

- ☐ Craft compelling meta descriptions that not only include keywords but also entice users to click through.
- ☐ Follow best practices for meta description length and format.

10. Page Speed

- ☐ Optimize your website's load speed by minimizing unnecessary scripts and using browser caching.
- ☐ Check your site's speed with tools like Google PageSpeed Insights and GTmetrix.

11. URL Structure

- ☐ Craft user-friendly, descriptive URLs that include keywords when applicable.
- ☐ Use hyphens (-) to separate words in URLs for better readability.

12. Schema Markup

- ☐ Implement schema markup to provide search engines with structured data about your content, making it more appealing in search results.

13. Internal Linking

- ☐ Implement internal links to connect related content within your website.
- ☐ Use descriptive anchor text that includes relevant keywords.

14. External Linking

- ☐ Include relevant external links to authoritative sources when appropriate, enhancing the credibility of your content.

15. Social Sharing

- ☐ Implement social sharing buttons on your content to encourage sharing and increase visibility.

16. HTTPS

- ☐ Secure your website with HTTPS to protect user data and improve search engine rankings.

17. Monitoring and Analysis

- ☐ Use tools like Google Analytics and Google Search Console to track the performance of your optimized pages.
- ☐ Adjust your strategy based on the insights gained from these tools.

18. User Intent

- ☐ Consider user intent when optimizing content. Ensure that your content matches what users are looking for when they search with specific keywords.

By diligently following this on-page optimization checklist, aviation training businesses can enhance their website's SEO, attract targeted traffic, and provide a better user experience, ultimately improving their online presence and achieving their marketing goals.

OFF-PAGE OPTIMIZATION

Backlinks from credible websites can significantly boost your SEO rankings. A backlink is a link from one website to another, also known as an inbound link. These links denote the relevance, quality, and authority of a web page to a search engine, leading to higher SERP rankings. A study by Backlinko confirms the number of domains linking to a page correlated with rankings more than any other factor.

OFF-PAGE OPTIMIZATION CHECKLIST FOR IMPROVED SEO

1. Backlink Strategy

- ☐ Develop a comprehensive backlink strategy aimed at acquiring high-quality and relevant backlinks from authoritative websites within the aviation industry.

2. Link Outreach

- ☐ Identify reputable aviation blogs, flight training forums, industry news websites, and other related platforms that could potentially link to your content.

3. Quality over Quantity

- ☐ Prioritize the quality of backlinks over their quantity. A few high-authority backlinks are more valuable than numerous low-quality ones.

4. Content Creation

- ☐ Produce high-quality and valuable content that naturally attracts backlinks from other websites. Content can include blog posts, guides, infographics, and industry reports.

5. Guest Blogging

- ☐ Consider guest blogging opportunities on aviation-related websites. Ensure that your guest posts are informative, well-researched, and provide value to the audience.

6. Link Building Outreach

- ☐ Reach out to webmasters, bloggers, and site owners to propose collaboration opportunities, such as guest posting, link exchanges, or content partnerships.

7. Broken Link Building

- ☐ Identify broken links on aviation-related websites and offer to replace them with relevant, high-quality content from your site.

8. Social Media Promotion

- ☐ Promote your content on social media platforms to increase its visibility and the likelihood of it being shared by others in the aviation community.

9. Online Forums and Communities

- ☐ Actively participate in aviation forums, communities, and discussion boards. Provide valuable insights, answers, and solutions while discreetly linking back to your site when relevant.

10. Monitor Mentions

- ☐ Use tools like Google Alerts or Mention to monitor online mentions of your brand or content. Engage with those who mention your site and potentially request backlinks when appropriate.

11. Directory Listings

- ☐ Ensure that your aviation training business is listed in relevant online directories and industry-specific listings.

12. Press Releases

- ☐ Publish press releases about noteworthy events or achievements related to your aviation training business. Distribute them through reputable channels to earn backlinks from news websites.

13. Collaborate with Influencers

- ☐ Partner with aviation influencers or experts in the field for co-authored content, interviews, or collaborations. Their endorsement can attract valuable backlinks.

14. Monitor Competitor Backlinks

- ☐ Analyze the backlink profiles of your competitors using tools like Ahrefs or SEMrush. Identify opportunities to acquire similar or better-quality backlinks.

15. Link Diversity

- ☐ Aim for a diverse range of backlinks from various sources, including blogs, news sites, forums, and educational institutions.

16. Avoid Black Hat Tactics

- ☐ Strictly avoid black hat SEO practices such as buying backlinks, link farms, or spammy link-building schemes. These can result in penalties from search engines.

17. Analyze Backlink Quality

- ☐ Regularly assess the quality and relevance of your existing backlinks. Disavow or request removal of low-quality or spammy backlinks that may harm your site's reputation.

18. Track Backlink Performance

- ☐ Monitor the impact of your backlink-building efforts on your site's rankings and organic traffic. Use tools like Google Search Console and SEO analytics platforms to track progress.

By following this off-page optimization checklist, aviation training businesses can systematically build a strong backlink profile, improve their SEO rankings, and establish themselves as authoritative entities within the aviation industry. Remember that the quality and relevance of backlinks are key, and a strategic, ethical approach to off-page optimization is essential for long-term success.

CONTENT MARKETING

Content is the foundation of SEO. The Content Marketing Institute found that small businesses with blogs get 126% more leads than small businesses without. Publish informative articles, how-to guides, and FAQs that answer common queries from potential students. Not only will this help with SEO, it will also position you as an authority in your field. We'll cover this more in the next chapter.

Enhancing Local SEO for Aviation Training Schools

For flight and ground schools, local SEO is crucial. Moz reports that 46% of all Google searches are local. Make sure you claim and regularly update your Google My Business account with photos, reviews, and relevant information. Schools should prompt students for reviews, whether an instructor hands them a card with a QR code and you send an automated email after a flight.

Mastering local SEO is essential to connect with potential students in your vicinity. In a digital age where local searches are increasingly prevalent, optimizing your online presence is paramount. Here's a comprehensive guide on how aviation training schools can improve their local SEO, leveraging Google, Bing, and Apple local searches.

CONTENT IS THE FOUNDATION OF SEO. POSITION YOURSELF AS AN AUTHORITY.

CONTENT MARKETING CHECKLIST

Google My Business (GMB) Optimization

- ☐ **Create or Claim Your Listing:** Start by creating a Google My Business account or claiming an existing one. Ensure all information is accurate and complete.
- ☐ **Accurate NAP:** Consistency is key; make sure your Name, Address, and Phone number (NAP) are consistent across all online platforms.
- ☐ **Categories and Attributes:** Select relevant categories and attributes that accurately describe your aviation training school.
- ☐ **Photos and Videos:** Regularly update your GMB profile with high-quality images and videos showcasing your facilities, instructors, and aircraft.
- ☐ **Customer Reviews:** Encourage satisfied students to leave reviews on your GMB listing. Respond to reviews, both positive and negative, to engage with your audience.
- ☐ **Posts and Updates:** Utilize the GMB Posts feature to share news, events, and updates about your school, helping to keep your listing active and engaging.

Bing Places for Business

- ☐ **Claim Your Bing Listing:** Create or claim your listing on Bing Places for Business, mirroring the information on your GMB profile.
- ☐ **Complete Your Profile:** Fill in all available fields, including a detailed description of your aviation training services.
- ☐ **Images and Videos:** Upload compelling visuals that highlight your school's unique offerings.
- ☐ **Customer Reviews:** Encourage students to leave reviews on your Bing listing, as positive reviews can boost your search visibility.

Apple Maps Connect

- ☐ **Listing Verification:** Sign into Apple Maps Connect and verify your listing through the verification code sent to your business phone number.
- ☐ **Accurate Information:** Ensure that all details, such as your school's name, address, and contact information, are accurate and up to date.
- ☐ **Categories:** Select appropriate categories that align with your aviation training services.
- ☐ **Imagery:** Upload eye-catching images that showcase your school's facilities, aircraft, and training environment.

LOCAL SEO CHECKLIST FOR AVIATION TRAINING SCHOOLS

1. Claim and Verify Listings

- ☐ Create or claim your listings on Google My Business, Bing Places for Business, and Apple Maps Connect.

2. Consistent NAP

- ☐ Maintain a consistent Name, Address, and Phone number (NAP) across all listings and directories.

3. Complete Profiles

- ☐ Fill out all available fields with accurate information, including a detailed description of your aviation training services.

4. High-Quality Visuals

- ☐ Regularly update your profiles with high-resolution images and videos that reflect the quality of your school.

5. Customer Reviews

- ☐ Encourage satisfied students to leave reviews on all platforms, and respond to reviews promptly.

6. Categories and Attributes

- ☐ Select relevant categories and attributes that accurately represent your aviation training school.

7. Regular Updates

- ☐ Keep your profiles active by posting news, events, and updates about your school.

8.Local Keywords

- ☐ Optimize your website and content for local keywords related to aviation training and your location.

9. Mobile Optimization

- ☐ Ensure your website is mobile-friendly, as many local searches occur on mobile devices.

10. Local Link Building

- [ ] Build local backlinks from aviation-related businesses and directories.

11. Monitoring and Analytics

- [ ] Regularly monitor your local SEO performance using tools like Google Analytics and Search Console.

12. Stay Informed

- [ ] Stay up to date with changes in local SEO algorithms and best practices to maintain and improve your rankings.

By diligently following this checklist and optimizing your presence on Google, Bing, and Apple Maps, your aviation training school can enhance its local SEO efforts, making it easier for potential students in the area to discover and choose you.

TECHNICAL SEO

Factors like website speed, mobile optimization, and secure, accessible websites (HTTPS) are important for search engines. Google's PageSpeed Insights can help you understand how to improve your site speed, which is a known ranking factor.

The Significance of Technical SEO: Elevating Your Website's Performance

In the ever-evolving realm of search engine optimization, technical SEO stands out as a crucial component that directly influences a website's visibility and ranking on search engine results pages (SERPs). This facet of SEO focuses on enhancing the technical aspects of a website, ensuring that it is not only user-friendly but also optimized for search engines. This section explains the importance of technical SEO and some key factors that can make a substantial difference.

Website Speed

One of the foremost technical SEO considerations is website speed. In an era where users expect instant access to information, slow-loading websites can deter visitors and lead to higher bounce rates. Importantly, website speed is also a known ranking factor in search algorithms. Google, for instance, rewards faster websites with improved search visibility. Tools like Google's PageSpeed Insights provide valuable on how to boost your site's speed. Recommendations may include optimizing images, reducing unnecessary scripts, leveraging browser caching, and more. By implementing these optimizations, you enhance user experience and bolster your search engine rankings.

Mobile Optimization

In an increasingly mobile-centric digital landscape where over half of all website traffic comes from mobile devices, mobile optimization is paramount. Search engines like Google prioritize mobile-friendly websites when determining rankings, particularly for mobile search results. Responsive web design, which adapts seamlessly to various screen sizes and devices, is a fundamental aspect of mobile optimization. Ensuring that your website is responsive guarantees that users have a consistent and accessible experience, regardless of whether they're browsing on a desktop, smartphone, or tablet.

Secure and Accessible Websites (HTTPS)

Website security is a critical concern for both users and search engines. Google, in particular, has placed significant importance on this aspect by giving preference to secure websites. A secure website is denoted by the HTTPS protocol, which encrypts data transmitted between the user's browser and the server. Beyond security, HTTPS also fosters trust among users. Websites without it may display a "Not Secure" warning in web browsers, which can deter visitors. To improve both security and SEO, it's essential to obtain an SSL certificate and transition to HTTPS.

Accessibility and SEO

Web accessibility is about ensuring that all users, including those with disabilities, can access and navigate your website seamlessly. While primarily a consideration for inclusivity and legal compliance, web accessibility also indirectly impacts SEO. An accessible website is more likely to rank well because it caters to a broader audience and provides a better user experience, which are factors search engines consider when determining rankings.

Technical SEO is a pivotal aspect of a comprehensive SEO strategy. It encompasses factors like website speed, mobile optimization, and secure, accessible websites (HTTPS) that are crucial for both user experience and search engine visibility. By proactively addressing these technical elements, website owners and SEO practitioners can improve their site's performance, enhance user satisfaction, and boost their rankings in search engine results pages. In a digital landscape where competition is fierce and user expectations are high, technical SEO is a key driver of online success.

ANALYTICS & MONITORING CAN INCREASE ANNUAL GROWTH RATES BY OVER 30%.

ANALYTICS AND MONITORING

Consistently monitor your rankings, traffic, and conversions using tools like Google Analytics and Google Search Console. Knowing what works will help you refine your SEO strategy over time. Companies that are insight-driven are growing at an average of more than 30% annually, according to a report by Forrester.

Data-Driven Success: The Power of Analytics and Monitoring in SEO

In the dynamic world of SEO, staying ahead of the curve and achieving lasting success requires more than just optimization techniques. It necessitates a continuous cycle of analysis and refinement. This is where the vital role of analytics and monitoring comes into play. By consistently tracking your rankings, traffic, and conversions using tools like Google Analytics and Google Search Console, you can unlock insights that will enable you to fine-tune and enhance your SEO strategy over time. A report from Forrester—revealing that companies prioritizing insight-driven decision-making are experiencing annual growth rates exceeding 30%—underscores the importance of this data-driven approach.

Understanding the SEO Metrics

1. Rankings

Regularly monitoring your website's rankings in search engine results pages (SERPs) is fundamental to assessing your SEO performance. Tools like Google Search Console provide valuable data on the keywords that drive traffic to your site and how your rankings change over time. By keeping a close eye on these metrics, you can identify which keywords are performing well and which may need additional optimization.

2. Traffic

Google Analytics offers a comprehensive view of your website's traffic sources, user behavior, and engagement metrics. It reveals where your visitors come from, which pages they visit, and how long they stay. This data is helps you understand the effectiveness of your SEO efforts and identify opportunities for improvement.

3. Conversions

Ultimately, the success of your SEO strategy is determined by conversions, which can include sign-ups, purchases, or any other desired actions on your website. Tracking conversions in Google Analytics allows you to measure the ROI of your SEO efforts and identify which channels and keywords are driving the most valuable traffic.

Iterative Refinement

Armed with this data, you can embark on a process of iterative refinement. Here's how it works:

1. Identify Successes and Failures

Analyze your data to pinpoint which SEO strategies and tactics are delivering positive results and which are falling short.

2. Adapt and Optimize

Use your insights to adapt and optimize your SEO strategy. For example, if you discover that certain keywords consistently drive high-converting traffic, consider expanding your content around those topics.

3. Experiment and Test

SEO is an ever-evolving field, and what works today may not work tomorrow. Experiment with new strategies, such as creating fresh content or building high-quality backlinks, and monitor their impact on your rankings and traffic.

4. Measure and Adjust

Continuously measure the effects of your adjustments and make further refinements as necessary. SEO is an ongoing process, and staying responsive to changes is key to maintaining and improving your search visibility.

The Competitive Edge: Insights-Driven Growth

As referenced earlier, Forrester report a compelling advantage for companies that prioritize data and insights in their decision-making processes. Harnessing the power of analytics and monitoring in your SEO strategy, gives you a competitive edge and sets the stage for sustainable growth. Whether you're a small aviation training school or a large corporation, your ability to adapt and refine your SEO efforts based on data-driven insights can make all the difference in achieving and surpassing your goals.

CHECKLIST FOR IMPLEMENTING A SEO STRATEGY

- ☐ **Keyword Research:** Use tools like Google's Keyword Planner or SEMrush to identify relevant keywords.
- ☐ **On-Page Optimization:** Modify meta titles, meta descriptions, and header tags to include target keywords.
- ☐ **Off-Page Optimization:** Earn backlinks from authoritative domains in the aviation and training sectors.
- ☐ **Content Marketing:** Consistently produce high-quality, original content that is relevant to your target audience.
- ☐ **Local SEO:** Optimize your Google My Business account and focus on local keywords.
- ☐ **Technical SEO:** Ensure website speed, mobile optimization, and secure, accessible websites.
- ☐ **Analytics and Monitoring:** Use tools like Google Analytics and Google Search Console for consistent monitoring and optimization.

TAKE CONTROL WITH SEO

By methodically implementing these SEO strategies, you can climb the ranks on search engines, attracting more of the right eyeballs and converting them into paying students. This tactical approach will help your flight or ground school outperform competitors and solidify its reputation as a go-to destination for aviation training.

7

LEARN HOW TO CRAFT YOUR CONTENT TO DRIVE TRAFFIC AND INCREASE GROWTH RATES

CHAPTER SEVEN

CONTENT MARKETING

MORE THAN JUST BLOGS

Content marketing is much more than just publishing blogs; it's about creating valuable, relevant content to attract a clearly defined audience and drive profitable customer action. Content marketing serves as a multi-channel approach encompassing various formats from videos to podcasts, infographics to webinars, and beyond. According to the Content Marketing Institute, content marketing costs 62% less than traditional marketing yet generates about 3 times as many leads. For flight and ground schools, leveraging various types of content improves SEO while creating an online brand identity that appeals to both student pilots and aviation professionals.

CONTENT MARKETING

TAKE CONTROL OF YOUR CONTENT

Blogging for Success: Unlocking the Potential of Content Marketing

BUSINESSES THAT BLOG GENERATE **55%** MORE VISITORS AND **67%** MORE LEADS.

While the title of this chapter may seem to diminish the significance of blogging in the digital marketing landscape, the reality is quite the opposite. Blogging remains a cornerstone of content marketing and holds immense potential for businesses of all sizes. As revealed by a HubSpot report, businesses that maintain an active blog witness a remarkable 55% increase in website visitors and a staggering 67% boost in leads. However, the key to reaping these rewards lies not in mere content creation but in delivering value and addressing the specific needs of your target audience. This chapter explores the art and science of blogging, encompassing posting frequency, distribution strategies, social media channel selection, and more, to help aviation training businesses harness the full potential of this powerful tool.

QUALITY OVER QUANTITY: THE ART OF BLOGGING

One of the cardinal rules of successful blogging is to prioritize quality over quantity. While it's true that regular posting can help keep your audience engaged and attract search engine attention, the content itself must be valuable, relevant, and informative. In the aviation training niche, this translates to creating content that addresses the concerns, questions, and needs of your prospective and current students.

Posting Frequency

The optimal posting frequency can vary depending on your resources and audience expectations. Aim for consistency. Whether it's one well-researched, comprehensive blog post per week or two per month, make sure you can maintain your chosen frequency. Avoid sacrificing quality for quantity.

Keyword Research

Conduct thorough keyword research to identify topics and phrases relevant to aviation training. Tools like Google's Keyword Planner can help you uncover trending keywords and queries in your niche. Integrate these keywords naturally into your blog content to improve its visibility in search results.

Diverse Content Types

Don't limit yourself to traditional written blog posts. Embrace diverse content formats like videos, infographics, podcasts, and case studies. Different audiences prefer different content types, so diversifying can help you reach a broader spectrum of potential students. Repurpose your blog content into different formats. For example, transform a blog post into a video or create an infographic summarizing key points. This approach extends the lifespan of your content and appeals to diverse audience preferences.

Audience-Centric Approach

Craft your content with a deep understanding of your target audience's pain points and interests. For aviation training schools, this might include topics like FAA regulations, pilot training processes, career paths in aviation, and safety guidelines. Addressing these concerns directly demonstrates your expertise and builds trust.

PROMOTION AND DISTRIBUTION: MAXIMIZING REACH

Creating exceptional content is just the first step. Effective distribution and promotion are equally vital to ensure your blog posts reach your intended audience.

PRIORITIZE **QUALITY OVER QUANTITY** TO KEEP YOUR CONTENT VALUABLE, RELEVANT, AND INFORMATIVE.

Email Marketing

Leverage your email subscriber list to notify your audience of new blog posts. Regular e-newsletters or updates with a summary and links to recent blog content can keep your subscribers engaged.

Social Media

Select the social media platforms that align best with your target audience. Aviation training businesses may find platforms like LinkedIn, Facebook, and Twitter particularly effective. Share your blog posts, engage with your audience, and foster discussions around your content.

Guest Posting

Collaborate with aviation-related websites, forums, or influencers to guest post and share your expertise. This can expand your reach to a broader audience interested in aviation training.

Engage and Interact

Encourage comments and discussions on your blog posts. Respond to comments promptly, fostering a sense of community around your content. Engagement not only keeps visitors on your site longer but can also improve your search engine rankings.

SUCCESS IN BLOGGING IS A JOURNEY, UTILIZE ALL OF THE TOOLS AVAILABLE TO YOU ALONG THE WAY.

MEASURING SUCCESS: ANALYTICS AND INSIGHTS

Blogging's true power lies in its ability to provide measurable results. To optimize your blogging strategy, you must leverage analytics and insights.

Google Analytics

Use Google Analytics to track website traffic, user behavior, and conversion rates stemming from your blog posts. Identify which posts perform best and understand your audience's preferences.

Conversion Tracking

Set up conversion tracking within Google Analytics to measure the specific actions taken by visitors who read your blog posts. This could include signing up for newsletters, requesting more information, or enrolling in courses.

Keyword Performance

Monitor how your blog posts rank for target keywords. Tools like Moz and SEMrush can help you track keyword rankings and identify opportunities for improvement.

Audience Feedback

Actively seek and analyze feedback from your audience. Conduct surveys or use social media polls to understand which topics resonate most with your readers. This insight can guide future content creation.

Blogging remains a potent tool for aviation training businesses seeking to enhance their online presence, attract prospective students, and establish themselves as authoritative voices in the industry. By prioritizing content quality, understanding your audience's needs, and implementing effective distribution strategies, you can harness the full potential of blogging to generate increased website traffic, leads, and ultimately, student enrollments. Remember that success in blogging is a journey, and regularly measuring and adapting your strategy will keep you on the path to sustained growth and success.

BLOGGING FOR AVIATION TRAINING BUSINESSES: A SUCCESS CHECKLIST

- ☐ **Quality Over Quantity:** Prioritize high-quality, informative content over frequent posting. Strive for consistency without sacrificing content quality.
- ☐ **Posting Frequency:** Choose a posting frequency that aligns with your resources and audience expectations, whether it's weekly, bi-weekly, or monthly.
- ☐ **Keyword Research:** Conduct thorough keyword research to identify trending topics and phrases relevant to aviation training. Incorporate these keywords naturally into your content.
- ☐ **Diverse Content Types:** Embrace a variety of content formats, such as videos, infographics, podcasts, and case studies, to cater to different audience preferences.
- ☐ **Audience-Centric Approach:** Craft content that addresses the specific needs, concerns, and interests of your target audience, including topics like FAA regulations, pilot training processes, and career paths in aviation.
- ☐ **Promotion and Distribution:** Maximizing Reach
- ☐ **Email Marketing:** Notify your email subscribers of new blog posts through regular newsletters or updates that summarize recent content.
- ☐ **Social Media:** Choose the most relevant social media platforms for your audience and share blog posts. Engage with your audience and foster discussions around your content.
- ☐ **Guest Posting:** Collaborate with aviation-related websites, forums, or influencers to guest post and expand your reach.
- ☐ **Content Repurposing:** Transform blog content into different formats, such as videos or infographics, to appeal to diverse audience preferences.
- ☐ **Engage and Interact:** Encourage comments and discussions on your blog posts. Respond promptly to foster a sense of community around your content.
- ☐ **Measuring Success:** Analytics and Insights
- ☐ **Google Analytics:** Use Google Analytics to track website traffic, user behavior, and conversion rates resulting from your blog posts.

- ☐ **Conversion Tracking:** Set up conversion tracking within Google Analytics to measure specific actions taken by visitors after reading your blog posts.
- ☐ **Keyword Performance:** Monitor how your blog posts rank for target keywords using tools like Moz and SEMrush.
- ☐ **Audience Feedback:** Actively seek and analyze feedback from your audience through surveys or social media polls to understand their preferences.

By following this comprehensive checklist, aviation training businesses can systematically plan and execute their blogging strategy, with a focus on delivering value, engaging their audience, and measuring success to continuously refine their approach.

VIDEOS

A study by Wyzowl found that 86% of businesses use video as a marketing tool, and 93% of marketers say it's a vital part of their strategy. Platforms like YouTube are excellent for long-form content, while Instagram and TikTok are better suited for shorter, more engaging clips.

Taking Flight with Video Marketing: Crafting a Winning Strategy for Your Flight School

With information consumption being driven by visual appeal and instant gratification, video marketing has soared to the forefront as a powerful tool for aviation training businesses, including flight schools. Whether you aim to showcase real-world flight simulations, share student success stories, or offer engaging tutorials, video content holds the potential to captivate your audience and elevate your brand. In this section, we will guide flight school owners through the process of creating a comprehensive video marketing strategy that resonates with their target audience and propels their aviation training business to new heights.

1. Define Your Objectives

Before embarking on your video marketing journey, it's essential to establish clear and measurable objectives. Consider what you aim to achieve through your video content. Are you looking to increase enrollment numbers, build brand awareness, or engage with potential students? Your objectives will guide your content creation and distribution strategy.

2. Know Your Audience

Understanding your target audience is paramount in crafting effective video content. Flight school owners should create buyer personas that outline the characteristics, needs, and preferences of their ideal students. Tailoring your video content to address these personas, will provide valuable insights and solutions to their aviation training aspirations.

3. Choose the Right Platforms

Selecting the appropriate platforms for sharing your video content is vital. Each platform caters to different audience demographics and formats. For instance:

- **YouTube:** Ideal for hosting long-form content such as in-depth flight tutorials, virtual tours of your training facilities, and student testimonials.
- **Instagram:** Perfect for sharing shorter, engaging clips that provide quick insights into your aviation training programs, behind-the-scenes glimpses, and instructor spotlights.
- **TikTok:** Suited for creating short, attention-grabbing videos that showcase the thrill of flight training, safety protocols, and unique aviation experiences.

Your choice of platform should align with your target audience's preferred channels and content consumption habits.

CONSIDER **WHAT YOU AIM TO ACHIEVE** THROUGH YOUR VIDEO CONTENT WHEN PLANNING THEIR CREATION.

4. Craft Compelling Content

The heart of any successful video marketing strategy lies in creating compelling, informative, and engaging content. Consider these content ideas for your flight school:

- **Real-World Simulations:** Provide viewers with a taste of the aviation training experience through immersive simulations, flight training sessions, and cockpit insights.
- **Success Stories:** Share testimonials and success stories from past and current students to build trust and inspire potential enrollees.
- **Mini-Tutorials:** Offer bite-sized aviation tutorials on topics like pre-flight checklists, navigation techniques, and aircraft operations.
- **Behind-the-Scenes:** Give your audience a peek behind the curtain by showcasing your training facilities, dedicated instructors, and aircraft fleet.
- **Safety Demonstrations:** Highlight the importance of safety in aviation training with informative videos on safety protocols, emergency procedures, and risk management.

5. Consistency is Key

Maintaining a consistent posting schedule is crucial for audience engagement and brand visibility. Create an editorial calendar that outlines your video release dates, content themes, and distribution strategy. Consistency not only keeps your audience engaged but also reinforces your flight school's commitment to quality training.

6. Optimize for SEO

Video SEO is essential to ensure your content reaches a wider audience. Optimize your video titles, descriptions, and tags with relevant keywords related to aviation training and your specific offerings. This will enhance the discoverability of your videos on platforms like YouTube.

7. Encourage Engagement

Engagement is a two-way street. Encourage your audience to like, share, comment, and subscribe to your video content. Respond promptly to comments and engage in meaningful conversations to foster a sense of community around your brand.

8. Measure and Analyze

Monitoring the performance of your video content is crucial for refining your strategy over time. Track metrics such as view counts, watch time, click-through rates, and engagement levels. Tools like YouTube Analytics provide valuable insights into viewer demographics and behavior, helping you tailor future content to better resonate with your audience.

9. Invest in Quality

While content is king, the production quality of your videos also matters. Invest in good equipment, including cameras, microphones, and editing software, to ensure your videos look and sound professional.

10. Adapt and Evolve

The digital landscape is constantly evolving. Stay abreast of industry trends and emerging technologies to keep your video marketing strategy fresh and relevant. Be ready to adapt to changes in audience preferences and new opportunities that arise in the aviation training sector.

In conclusion, video marketing is a potent tool that flight school owners can leverage to connect with their audience, showcase their expertise, and promote their aviation training programs effectively. By crafting a well-defined strategy that aligns with your objectives, understanding your audience, and creating engaging, informative content, you can take flight into the world of video marketing and soar to new heights in the aviation training industry.

PODCASTS

Podcasting is another rapidly growing medium. Edison Research reports, the number of Americans who have listened to a podcast grew by 75% between 2014 and 2020. Flight schools can utilize this medium to discuss aviation news, interview experts, and explore subjects like the science of flight, aviation safety, and career development.

Taking Flight with Podcast Marketing: Crafting a Strategy for Your Flight School

As a flight school owner, podcasting presents a golden opportunity to engage with your audience, share industry insights, and establish your aviation training business as an authoritative voice in the field. In this section, we will guide you through the process of creating a comprehensive podcast marketing strategy tailored to your flight school's objectives and aspirations.

1. Define Your Podcasting Objectives

Before you embark on your podcasting journey, it's important to establish clear and measurable objectives. Consider what you aim to achieve through your podcast. Are you looking to build brand authority, engage with potential students, or showcase your flight school's expertise? Your objectives will serve as the compass for your content creation and distribution strategy.

2. Know Your Podcasting Audience

Understanding your target audience is paramount in crafting effective content. As a flight school owner, you already have a deep understanding of your prospective students, which you'll use to create detailed target personas as we discussed in Chapter 2. Be sure you tailor all of your content—including podcasts—to address the unique needs and preferences of these personas.

3. Choose the Right Podcast Format and Content

Podcasts come in various formats, each catering to different audience preferences. The essence of any successful podcasting strategy lies in creating engaging, informative, and entertaining content. Consider the following formats:

- **Interviews:** Host aviation experts, instructors, and successful alumni to share their insights and experiences.
- **Educational:** Dive deep into subjects like aviation safety, flight science, career development, and FAA regulations.
- **News and Updates:** Keep your audience informed about the latest industry news, flight training developments, and your school's updates.

- **Storytelling:** Share compelling stories of student achievements, memorable flights, and aviation adventures.
- **Q&A Sessions:** Address listener questions and concerns related to aviation training, safety, and careers.

The chosen format should align with your audience's interests and provide value.

4. Consistency and Scheduling

Maintaining a consistent podcasting schedule is crucial for building and retaining your audience. Create a podcasting editorial calendar that outlines your episode release dates, content themes, and distribution strategy. Consistency reinforces your flight school's commitment to quality training and content.

5. Invest in Quality

Production quality plays a significant role, since podcasts—like all of your other marketing tactics—are a direct reflection on your brand. Invest in quality audio recording equipment, including microphones and editing software, to ensure your podcast sounds professional and engaging.

6. Optimize for SEO

Podcast SEO is essential to ensure your content reaches a broader audience. Optimize your podcast titles, descriptions, and show notes with relevant keywords related to aviation training and your specific offerings.

7. Encourage Engagement

Engagement is vital for building a loyal podcast audience. Encourage your listeners to submit questions, share their experiences, and provide feedback. Interact with your audience through social media, emails, and live Q&A sessions.

8. Measure and Analyze

To refine your podcasting strategy over time, monitor the performance of your episodes. Track metrics such as episode downloads, listener demographics, and engagement levels. Analyze this data to understand what resonates most with your audience.

PODCASTING IS A DYNAMIC AND RAPIDLY GROWING CONTENT MEDIUM.

9. Promotion and Distribution

Effective promotion and distribution are key to expanding your podcast's reach. Utilize various channels:

- **Podcast Directories:** Submit your podcast to popular directories like Apple Podcasts, Spotify, Google Podcasts, and Stitcher.
- **Social Media:** Share teaser clips, episode announcements, and behind-the-scenes content on your social media platforms.
- **Email Marketing:** Notify your subscribers when a new episode is released.
- **Collaborations:** Collaborate with other aviation-related podcasts, forums, or influencers to cross-promote your content.

10. Adapt and Evolve

Stay attuned to industry trends and emerging technologies to keep your podcasting strategy fresh and relevant. Be ready to adapt to changes in listener preferences and new opportunities in the aviation training sector.

Podcast marketing offers flight school owners a dynamic and engaging platform to connect with your audience, share industry insights, and promote their aviation training programs effectively. By creating a well-defined strategy that aligns with your objectives, understanding your audience, and producing engaging, informative content, you can easily integrate podcasting into your marketing mix.

INFOGRAPHICS

According to the Social Media Examiner, infographics are liked and shared on social media 3 times more than other any other type of content. They can help simplify complex aviation topics like flight mechanics, air traffic control procedures, or aviation history, making the information more digestible for your target audience.

Creating effective infographics is essential for engaging your audience and conveying intricate aviation topics clearly. Here are some best practices:

- **Clarity and Simplicity:** Keep your infographics clean and uncluttered. Use a logical flow and concise language to simplify complex concepts. The goal is to make information easy to understand at a glance.
- **Visual Appeal**: Use a cohesive color scheme, high-quality images, and visually pleasing design elements that all adhere to your brand guidelines. Visual appeal can grab your audience's attention and keep them engaged.
- **Storytelling:** Organize your information in a narrative format, guiding the viewer through a story or process. This helps make the content relatable and memorable.
- **Data Visualization:** If your infographic includes data, use charts or graphs to present it visually. Visual representations of data can be more compelling than numbers alone.
- **Mobile-Friendly:** Ensure your infographic is responsive and looks good on various devices, especially mobile phones, as many users access content on smartphones.
- **Distribution:** Leverage your infographics across your content - blogs, social media, email newsletters, and videos to maximize the benefit.
- **Shareability:** Make it easy for viewers to share your infographic on social media platforms. Include share buttons and a compelling call-to-action. Don't forget to add a watermark so your school receives proper credit and a link back to your content.
- **Cite Sources:** If you're using data or statistics, provide proper citations to establish credibility and transparency.
- **Testing:** Test your infographic with a sample audience to gather feedback and make improvements before widespread distribution.

By following these best practices, you can create infographics that resonate with your aviation training audience, simplify complex topics, and encourage social media sharing, ultimately enhancing your brand's reach and engagement.

Creating infographics can be achieved through two primary approaches: using your own tools or hiring freelancers. If you opt for the DIY route, there are various online infographic maker tools available, such as Canva, Piktochart, and Visme, which offer user-friendly templates and design elements to help you create visually appealing infographics. These tools often require little to no design experience.

If you prefer a professional touch or lack the time to design infographics yourself, hiring freelancers can be an excellent option. Freelance platforms like Upwork, Freelancer, and Fiverr connect you with experienced graphic designers and infographic creators. You can browse through portfolios, assess freelancer reviews, and select a skilled individual who aligns with your project's requirements and budget. Freelancers bring expertise in design, ensuring that your infographics are not only visually appealing but also effectively convey your aviation training messages. Whether you choose to go the DIY route or hire freelancers, infographics can help simplify complex aviation topics and engage your target audience.

INFOGRAPHICS ARE LIKED AND SHARED ON SOCIAL MEDIA **3 TIMES MORE** THAN ANY OTHER TYPE OF CONTENT.

WEBINARS

Webinars are highly effective for B2B marketing; they can also be beneficial in an educational context. Capterra notes that 73% of B2B marketers and sales leaders say a webinar is the best way to generate high-quality leads. Flight and ground schools can offer live Q&A sessions, training previews, or even guest lectures by aviation experts.

Creating a webinar strategy aligned with your defined marketing strategy is a strategic move for aviation training businesses looking to engage their audience and build industry authority. Here's a step-by-step guide for creating an effective webinar strategy:

1. Define Your Webinar Objectives

Start by clearly outlining the objectives of your webinar. Your goals might include generating leads, showcasing your expertise, increasing brand visibility, or educating potential students about aviation training programs.

EXAMPLES OF SUCCESSFUL INFOGRAPHICS

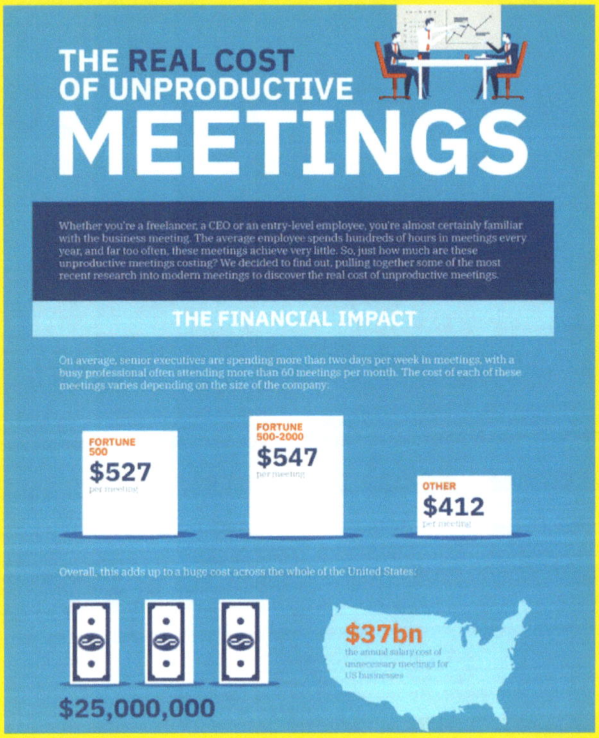

Images: https://reallygooddesigns.com/infographic-examples/

CREATE YOUR WEBINAR CONTENT BASED ON YOUR MARKETING STRATEGY.

2. Know Your Audience

Leverage the buyer personas you've already defined in your marketing strategy, by tailoring your webinar content to address the needs and pain points of your target audience.

3. Select Relevant Topics

Choose webinar topics that align with your aviation training business's core offerings and your audience's interests. For instance, you could host webinars on FAA regulations, pilot training tips, career prospects, or safety protocols.

4. Design Engaging Content

Create compelling and informative presentations or discussions for your webinar. Incorporate visuals, data, and real-world examples to keep your audience engaged throughout the session.

5. Choose the Right Format

Webinars can take various formats, including solo presentations, panel discussions, Q&A sessions, or live demonstrations. Select the format that best suits your content and audience engagement goals.

6. Set a Schedule

Determine the frequency of your webinars. Whether you opt for monthly, bi-monthly, or quarterly webinars, consistency is vital. Make sure to schedule webinars at times convenient for your target audience.

7. Promote Effectively

Leverage your marketing channels, including your website, social media, email marketing, and blog, to promote your upcoming webinars. Use enticing visuals, compelling messaging and clear CTAs to attract registrations.

8. Choose a Platform

Select a reliable webinar platform or software that allows you to host, manage, and track your sessions seamlessly. Platforms like Zoom, GoToWebinar, and Webex are popular choices.

9. Practice and Rehearse

Before going live, rehearse your webinar multiple times to ensure a smooth delivery. Familiarize yourself with the platform's features and troubleshoot any technical issues in advance.

THE RIGHT CONTENT CAN **SUPERCHARGE** YOUR SOCIAL MEDIA PRESENCE.

10. Engage Your Audience

During the webinar, encourage audience interaction through polls, chat, and Q&A sessions. Address questions and engage with participants to create a dynamic and informative experience.

11. Provide Value

Deliver valuable insights, actionable takeaways, and exclusive offers or discounts to attendees. Ensure that your audience feels they have gained something valuable from the webinar.

12. Follow-Up and Nurture

After the webinar, follow up with attendees by sending thank-you emails, webinar recordings, and additional resources. Continue to nurture these leads through targeted email marketing campaigns.

13. Measure and Optimize

Analyze key performance metrics, such as registration rates, attendance rates, engagement levels, and conversion rates. Use these insights to refine your webinar strategy and align it with your overall marketing strategy.

By integrating webinars into your aviation training business's marketing strategy, you can engage your audience on a deeper level, showcase your expertise, and drive leads and conversions. Remember that successful webinars are an ongoing process of planning, execution, and refinement, and they can be a powerful tool for achieving your marketing objectives.

SOCIAL MEDIA POSTS

The right content can supercharge your social media presence. Different platforms attract different demographics, so choose the ones your target personas are most likely to use, and tailor your content accordingly. LinkedIn is excellent for professional and career-related posts, while Instagram and Twitter can be more casual and visually engaging. According to Sprout Social, 45% of consumers would consider turning to social media for customer service issues, so having a strong content strategy can even help with customer retention. We'll dive into this in "Chapter 8: Social Media Marketing - Building a Community of Aviators."

E-BOOKS AND WHITEPAPERS

If you're looking for ways to share in-depth information on aviation topics, offering e-books or whitepapers can be an excellent strategy. These resources provide value and can also be used to capture leads. Marketo states that long-form content like e-books is particularly effective for lead generation, which 50% of marketers use for that purpose.

Unlocking the Power of E-Books and Whitepapers in Your Flight School's Marketing Strategy

E-books and whitepapers hold the potential to be powerful assets within your flight school's marketing arsenal. Here's how to create and leverage e-books and whitepapers as a crucial part of your aviation training marketing strategy:

1. Identify Your Target Audience

Start by defining your target audience, understanding their interests, needs, and pain points related to aviation training. Tailoring your e-books and whitepapers to address these specifics ensures they resonate with your potential students.

2. Choose Relevant Topics

Select topics that align with your training programs and require more in-depth explanations, such as "The Comprehensive Guide to Becoming a Pilot," "Navigating FAA Regulations for Beginners," or "Advanced Flight Maneuvers: A Practical Guide."

E-BOOKS AND WHITEPAPERS ARE AN INVALUABLE **LEAD GENERATION TOOL.**

3. In-Depth Research

Conduct thorough research to gather accurate and up-to-date information for your e-books and whitepapers. Cite credible sources and provide references to enhance the content's credibility.

4. Engaging Content Structure

Organize your content in a clear and engaging structure. Use headings, subheadings, bullet points, and visuals like images, charts, and infographics to break down complex information and make it easily digestible.

5. High-Quality Writing

Ensure your content is well-written, free of errors, and maintains a professional tone. If necessary, consider hiring a skilled writer or editor to polish your content.

6. Design and Branding

Design an attractive cover and layout for your e-books and whitepapers that align with your flight school's branding. Visual appeal can significantly impact the perceived value of your resources.

7. Add Value and Insights

Offer valuable insights, practical tips, and actionable advice within your content. Provide solutions to common challenges potential students might face during their aviation training journey.

8. Lead Generation

To capture leads, include strategically placed CTAs throughout your e-books and whitepapers. Invite readers to subscribe to your newsletter, request more information, or schedule a consultation.

9. Distribution Channels

Promote your e-books and whitepapers. Share them on your website, blog, and social media platforms. Consider using landing pages dedicated to each resource to optimize lead capture.

10. Email Marketing Integration

Integrate your e-books and whitepapers into your email marketing campaigns. Use targeted email lists to deliver these resources to the right audience segments.

11. Analytics and Optimization

Track the performance of your e-books and whitepapers using analytics tools. Monitor metrics such as downloads, leads generated, and engagement levels. Use these insights to refine your content and promotion strategy.

12. Follow-Up and Nurturing

After leads are captured, establish a nurturing process. Follow up with valuable content, educational materials, and personalized communication to guide potential students through the enrollment journey.

EMAIL IS **NEARLY 40 TIMES MORE EFFECTIVE** AT ACQUIRING NEW CUSTOMERS THAN FACEBOOK AND TWITTER COMBINED.

Incorporating e-books and whitepapers into your aviation training marketing strategy can offer valuable resources to your audience and effectively capture and nurture leads into enrolled students. These long-form content assets have the potential to establish your flight school as an authority in the field while driving engagement and conversions.

E-NEWSLETTERS

E-newsletters are a traditional yet highly effective method of content marketing. A McKinsey study found, email almost 40 times more effective at helping businesses acquire new customers than Facebook and Twitter combined. For flight schools, e-newsletters could feature updates, student success stories, and upcoming courses.

Harnessing the Power of E-Newsletters in Aviation Training Marketing

E-newsletters are a time-tested and valuable marketing tactic for flight schools to connect with their audience. They offer an avenue to keep your audience informed, engaged, and motivated. In the next chapter, we'll discuss the specifics of crafting compelling e-newsletters for aviation training marketing, for now, let's explore why this strategy remains indispensable.

E-NEWSLETTERS ARE A TRADITIONAL **YET HIGHLY EFFECTIVE** METHOD OF CONTENT MARKETING.

1. Direct Line of Communication

Unlike social media platforms, where algorithms can limit your reach, e-newsletters land directly in your subscribers' inboxes, ensuring your message is delivered. This establishes a direct and personal line of communication with your potential and current students.

2. Timely Updates

Flight schools can leverage e-newsletters to provide timely updates to their audience. Whether sharing the latest course offerings, announcing important industry developments, or providing insights into aviation regulations, e-newsletters keep your audience in the loop.

3. Showcasing Success Stories

One of the most compelling elements of aviation training is the success stories of your students. Newsletters offer a platform to showcase these stories, illustrating how your training programs have transformed aspiring aviators into skilled pilots. These stories inspire trust and confidence in your school.

4. Engaging Content

E-newsletters aren't limited to text alone. They provide the canvas to share a variety of content types, including videos, images, infographics, and links to in-depth articles. This diversity keeps your e-newsletters engaging and visually appealing.

5. Personalization

Email marketing platforms offer sophisticated tools for personalization. Flight schools can segment their email lists based on factors like training preferences, progress, or location, allowing for tailored content that resonates with each group.

SEGMENT YOUR EMAIL LIST TO **TAILOR CONTENT THAT RESONATES** WITH EACH GROUP.

6. Lead Generation and Conversion

Strategically placed CTAs within your e-newsletters can drive lead generation and conversion. Encourage subscribers to click through to enroll in courses, schedule consultations, or explore additional training options, turning readers into paying students.

7. Analytics and Optimization

Email marketing platforms provide comprehensive analytics that help flight schools track the performance of their e-newsletters. Metrics such as open rates, click-through rates, and conversion rates offer valuable insights for continuous optimization.

We'll provide more detail on crafting compelling e-newsletters tailored specifically for aviation training marketing in the next chapter. From optimizing subject lines to designing visually appealing templates, we'll share the secrets to creating newsletters that not only resonate with your audience but also drive results.

CONTENT MARKETING CHECKLIST

- ☐ **Blogs:** Regularly publish value-driven SEO-optimized article.
- ☐ **Videos:** Use various platforms to distribute different types and formats of video content.
- ☐ **Podcasts:** Establish a regular podcast to engage your audience.
- ☐ **Infographics:** Utilize graphics to simplify complex topics.
- ☐ **Webinars:** Host regular webinars.
- ☐ **Social Media Posts:** Maintain an active presence across relevant social channels.
- ☐ **E-books and Whitepapers:** Offer downloadable, value-driven long-form content.
- ☐ **E-newsletters:** Use email marketing to keep your audience updated and informed.

DIVERSIFY YOUR CONTENT MARKETING

Content marketing is not one-size-fits-all; it requires a diversified approach that caters to various consumer preferences and platforms. This multi-channel strategy, when executed effectively, can significantly boost your flight or ground school's visibility and credibility, helping to differentiate your brand in a competitive market.

8

UNDERSTAND THE POWER OF EMAIL MARKETING FOR ENGAGEMENT AND GROWTH

CHAPTER EIGHT

EMAIL MARKETING

KEEP THEM ENGAGED BEYOND THE COCKPIT

In the digital era, where instant messaging and social media reign supreme, you might wonder if email marketing still belongs in your a schools marketing strategy. The answer is an unequivocal yes. There were over 4 billion email users worldwide in 2020, and Statista expects that number to grow to 4.6 billion by 2025. Email marketing remains one of the most effective ways to reach and engage an audience. For flight and ground schools, it's an indispensable tool for maintaining continuous engagement with both aspiring and professional pilots. Email marketing can provide your audience with valuable resources, updates, course details, and other content tailored to their aviation aspirations.

A McKinsey report found that email marketing is nearly 40 times more effective at acquiring new customers than Facebook and Twitter combined. Moreover, the ROI for email marketing can be substantial. According to a study by VentureBeat, email marketing has an ROI of $38 for every $1 spent, making your email list an invaluable asset.

EMAIL MARKETING

UTILIZE AND GROW YOUR EMAIL LIST

In the aviation training context, email serves multiple functions, it can be both a promotional channel and an educational one. Through carefully segmented email lists, schools can send targeted information that helps students at different stages of their training, from ground school theory to advanced flight maneuvers. This channel also serves to build trust and brand loyalty, turning one-time students into lifelong advocates for your aviation training programs.

Given these compelling statistics and functions, email should be a cornerstone in your flight or ground school's digital marketing strategy. This chapter will delve into best practices for developing an email marketing strategy that will boost enrollment and foster a community of passionate aviators.

SEGMENTATION

SEGMENTED EMAIL CAMPAIGNS CAN INCREASE REVENUE BY **760%**

One of the biggest mistakes in email marketing is the "one-size-fits-all" approach. Marketers have noted a 760% increase in revenue from segmented campaigns compared to non-segmented campaigns, according to a study by Campaign Monitor. For flight schools, segmentation can be based on various factors such as the level of training, types of courses interested in, or even career aspirations in aviation.

Tips on Segmentation for Optimal ROI

Effective email segmentation starts with collecting the right kind of data about your subscribers. Flight and ground schools should collect information like the student's level of flight training (beginner, intermediate, advanced), type of license they're pursuing (private pilot, commercial pilot, flight instructor), and even career aspirations (commercial airline, cargo, general aviation, etc.). In addition, gather data on their interaction with past emails—did they click on a particular course link or download a training e-book? All these data points will help you send highly targeted emails that resonate with specific audience segments.

Segmentation is not just about splitting your email list into smaller parts; it's about understanding the unique needs and behaviors of these subgroups. You can even go as far as segmenting by geographical location, ideal for flight and ground schools that have multiple training locations or offer specialized regional training courses.

Here are some tips for effective segmentation:

- **Behavioral Segmentation:** Emails triggered by specific user behaviors—such as clicking on a 'Learn More' link about a particular training program—can yield exceptionally high open and conversion rates.

- **Stage-Specific Content:** Create different email sequences for students at various stages of training. Early-stage students may benefit from content that covers the basics, while advanced students will appreciate in-depth material or updates on additional certification programs.

- **Personalization:** Use the information you've gathered to personalize not just the greeting but also the content. According to a study by Experian, personalized emails can deliver 6 times higher transaction rates.

- **Retargeting:** If a prospective student has browsed your course offerings but hasn't enrolled yet, a retargeting email sequence could nudge them to complete their registration.

SEGMENTED EMAIL CAMPAIGNS HAVE **14.31% HIGHER OPEN RATES AND 100.95% HIGHER CLICK-THROUGH RATES** THAN NON-SEGMENTED CAMPAIGNS.

How Segmentation Can Help Growth

Segmentation can significantly improve the impact of your email marketing efforts. By targeting specific groups within your larger list, you're delivering highly relevant content that appeals directly to those recipients. This increases open rates, click-through rates, and customer engagement, strengthening your relationship with each subscriber.

A study by Mailchimp found that segmented email campaigns had 14.31% higher open rates and 100.95% higher click-through rates than non-segmented campaigns. These improved metrics can translate directly into increased enrollments and revenue for your flight or ground school. Moreover, segmentation can reduce the number of unsubscribes and spam complaints, thereby enhancing the longevity and ROI of your email list. In the long run, this contributes to steady growth, helping you to build a loyal community of students and aviation professionals.

IN THE REALM OF EMAIL MARKETING FOR FLIGHT AND GROUND SCHOOLS, CONTENT IS YOUR LINCHPIN. HOWEVER, NOT JUST ANY CONTENT WILL DO. THE KEY IS TO OFFER A DIVERSE YET TARGETED RANGE OF CONTENT THAT ADDRESSES THE NEEDS, ASPIRATIONS, AND QUESTIONS YOUR AUDIENCE MIGHT HAVE.

CONTENT

When it comes to email marketing for flight and ground schools, content is the fuel that drives engagement, conversions, and ultimately, revenue. Crafting high-quality content goes beyond sending out periodic newsletters or updates. It's about delivering value in a format that resonates with your specific audience segments. Below are some content strategies and tips that can help flight and ground schools achieve optimal ROI on their email campaigns:

Flight and ground schools, content is the linchpin in your email marketing. However, not just any content will do. The key is to offer a diverse yet targeted range of content that addresses the needs, aspirations, and questions your audience might have. Let's take a journey through various types of content that can elevate your email campaigns from mundane to magnificent.

Start with educational content, your ticket to positioning your institution as a thought leader in the field. Imagine sending out an email that discusses flight training intricacies or spells out the safety protocols in aviation. You're not just filling their inbox; you're adding value to their training journey, increasing their confidence in your institution's expertise.

But you're not only an educator; you're also an inspirer. Share stories of accomplished alumni or instructors who make flying less daunting for the novices. This is your evidence, your compelling case that assures them, "Studying with us will take you places."

There's something inherently alluring about being part of an 'inner circle,' which is why exclusive content holds such appeal. Capitalize on this human tendency by sending out exclusive offers and incentives only to your email subscribers. Imagine the rush a potential student would feel upon receiving a limited-time offer to sign up for a course at a discount—a rush that often leads to action.

Who says emails should only be read? Why not make them interactive? Embed polls, quizzes, or infographics that not only entertain but provide you with actionable data. This two-way interaction makes the recipient an active participant, not just a passive consumer of your message.

And don't forget about the community aspect. Announce events like webinars and open houses through targeted emails. These events aren't just opportunities for students to learn more; they're also showcases for your school's offerings, and emailing ensures you reach a committed audience.

Let's talk about the power of validation. A testimonial or a positive review is often more effective than your own brand communications. It's the age of social proof, and according to BigCommerce, 88% of people trust user reviews as much as personal recommendations.[18] Be sure to dedicate a section in your email templates to feature these validations.

Multimedia elements can add a new layer to your emails. According to Forbes, video can increase click-through rates by as much as 200-300%. So rather than stick to text alone, enhance comprehension and engagement through video or audio content.

Every piece of content should culminate in a compelling call to action. It's not just about imparting information; it's about directing your reader toward the next step in their journey, whether that's requesting more information, signing up for a course, or attending a webinar.

We've discussed the importance of relevance throughout this book, and email makes it particularly easy to distribute relevant content. Segmentation allows you to tailor your emails according to where the student is in their journey, specific interests they've shown, or past actions they've taken. This customization doesn't just improve open rates; it significantly enhances user engagement.

In a world inundated with content, quality triumphs over quantity. It's far more beneficial to send out one superbly crafted, highly-targeted email per month than to bombard your list with generic content weekly. The key lies in creating content that enriches, enlightens, and encourages your readers to take the next step in their aviation journey.

Quality content drives engagement and, in turn, revenue. Segmented and targeted emails alone are responsible for generating 58% of all revenue, per HubSpot. By following these tips and keeping your audience's needs and interests at the forefront, you can optimize the ROI of your email marketing campaigns substantially.

ANALYTICS AND PERFORMANCE METRICS

Analytics are the backbone of any successful email campaign, particularly for specialized sectors like flight and ground schools. Simply sending out emails isn't enough; you must understand how they are performing and why. By utilizing analytics, schools can optimize their email lists to achieve the best possible ROI.

Start by examining your open rates. If you find these numbers lagging behind the industry standard—about 23% for education-related emails according to Litmus—it's time to scrutinize your subject lines. Subject lines are your first touchpoint with the audience, so they need to be compelling. A popular strategy to refine them is to engage in A/B testing by sending two versions of your email—each with a different subject line—to a portion of your audience. The subject line that performs better in terms of open rates can then be used for the rest of the audience. We'll discuss this further in the next section.

Next, turn your attention to click-through rate (CTR). A low CTR is often indicative of a disconnect between your email content and what your audience finds valuable or compelling. To improve this metric, review your email's body and call-to-action. Does your content sufficiently engage the reader to prompt an action? Are you providing clear next steps for them to take? Again, A/B testing is invaluable here, but this time apply it to different versions of your content and CTAs. Through iterative testing, you can incrementally improve your CTR.

HIGH OPEN RATES BUT **LOW CONVERSION RATES** ARE OFTEN DUE TO PROBLEMS WITH YOUR LANDING PAGE.

Conversion rate tells you the percentage of email recipients who complete the action you desire, whether that's signing up for a course, downloading a resource, or making an inquiry. If your emails have high open rates and CTRs but low conversion rates, the issue often lies with the landing page your email links to. Ensure that these pages are streamlined, free of distractions, and contain a strong, clear CTA. Essentially, your landing page should be an extension of your email, continuing the conversation and guiding the user toward conversion.

Another important email metric is your bounce rate. Emails can bounce for a myriad of reasons, and each bounce represents a lost opportunity for engagement and conversion. Your analytics will show whether the bounces are 'hard' or 'soft', which can be indicative of email list health issues or more easily remediable issues like server downtime, respectively.

Consider analytics to be the co-pilot of your email marketing strategy, particularly in specialized fields like flight and ground schools. When utilized effectively, these metrics offer actionable data that amplifies your ROI. Each metric serves as a key performance indicator, from open rates and CTRs to conversion and bounce rates. Continual tracking, analyzing, and optimization in response to these metrics can turn your email marketing from a mere communication tool into a powerful engine for growth and profitability.

In a data-driven era, ignoring analytics is like flying blind. Understanding your email metrics doesn't just offer insight; it provides actionable data that can substantially improve your email marketing ROI.

ANALYTICS SHOULD BE CONSIDERED **THE CO-PILOT** OF YOUR EMAIL MARKETING STRATEGY.

A/B TESTING

As we mentioned in the previous section, A/B testing involves sending two versions of an email to see which performs better. Whether it's the subject line, content, or CTA, A/B testing allows for data-driven decisions that can improve conversion rates by up to 49%, according to OptinMonster.

A/B testing serves as a powerful technique for fine-tuning your email campaigns. The concept is simple but extraordinarily effective: you send out two versions of an email—each with a different element you want to test—to a subset of your audience. It could be as straightforward as changing the subject line or as complex as reformatting the entire email layout. Once both versions are sent, you assess which one outperforms the other based on predetermined metrics like open rate, click-through rate, or conversion rate.

The beauty of A/B testing lies in its capacity for continuous improvement. Once you've gathered your data and identified the winning version, you can use that as the new standard. From there, you can initiate another A/B test to refine yet another element, content, images, or calls to action. This iterative process facilitates a cycle of perpetual enhancement, allowing your email campaigns to evolve and adapt in real-time. In specialized sectors like flight and ground schools, where specific messaging can make a significant difference in enrollment rates and student engagement, this capacity for precise, data-driven adjustments is invaluable.

A/B testing also contributes to a more personalized user experience. By understanding what resonates with your audience, you can tailor future campaigns to better meet their preferences and needs, thus increasing the likelihood of engagement and conversion. For instance, if you find that emails featuring alumni success stories generate more clicks and conversions, that's an indicator that your audience values real-world outcomes and testimonials. You can then prioritize such content in future emails to increase relevance and engagement.

By employing A/B testing, you're not just guessing what might work; you're making informed, data-driven decisions that can significantly impact your ROI. As we move further into an age where data is plentiful, but attention spans are limited, optimizing every interaction with your audience will be key to maintaining a competitive edge. Therefore, integrating A/B testing into your email marketing strategy isn't just advisable; it's essential.

AUTOMATION

Email automation can save time and provide a personalized experience for your subscribers. For example, you could set up an automated series of welcome emails for new students. A study by Lenskold Group found that 63% of companies that are outgrowing their competitors use marketing automation.

A/B TESTING CAN LEAD TO CONVERSION RATE IMPROVEMENTS OF UP TO **49%**

Email automation is not just a tool for convenience; it's a powerful engine that can substantially impact your bottom line. Automating your email marketing efforts saves time while considerably boosting ROI by enabling highly targeted, timely, and personalized communications with your audience.

The power of email automation is also reinforced by findings from VentureBeat, which noted that automated emails generate 320% more revenue than non-automated emails. This surge in revenue can be attributed to the higher engagement rates automated emails typically receive due to their relevant and timely nature. When students sign up for a course or make an inquiry, automated welcome emails can instantly nurture this fresh interest, setting the stage for a deeper, more profitable relationship.

Moreover, automation allows for sophisticated segmentation strategies that can impact your bottom line. According to Campaign Monitor, marketers who used segmented campaigns reported as much as a 760% increase in revenue. Your flight school could set up different tracks for student pilots and aviation professionals, thereby serving each segment with content that resonates most with them. These automated, segmented emails keep your audience engaged, leading to higher open rates, click-through rates, and ultimately, conversion rates.

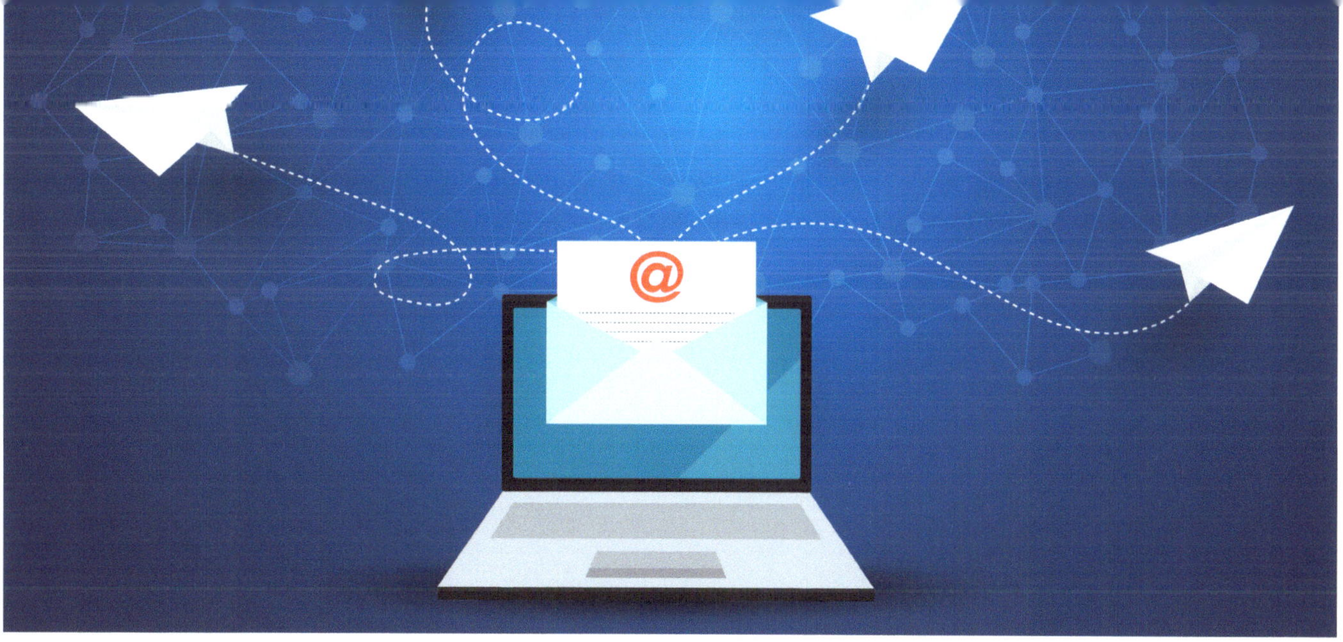

The value of email automation extends beyond student acquisition. You can also improve retention metrics by setting up automated emails for various stages in the customer lifecycle. Automating follow-up emails post-training to solicit feedback or to send reminders for license renewals can maintain ongoing relationships that could result in future enrollments or referrals.

To sum it up, the strategic deployment of email automation is a game-changer for any business, but especially for specialized sectors like flight and ground schools. It enhances engagement, promotes personalized interactions, and most critically, leads to a substantial improvement in ROI. With all these advantages, it's easy to see why companies that leverage marketing automation are the ones that usually outgrow their competition.

LEGAL COMPLIANCE

If your business uses email, your compliance with legal guidelines like the CAN-SPAM Act in the US or the General Data Protection Regulation (GDPR) in the European Union is non-negotiable. The stakes are high; failure to comply can result in severe financial penalties and significant damage to your brand's reputation. According to the Federal Trade Commission (FTC), the penalty for each separate email in violation of the CAN-SPAM Act is up to $43,280. On the GDPR front, non-compliance can lead to fines of up to €20 million or 4% of a company's annual global turnover, whichever is higher.

Aside from the financial repercussions, compliance also impacts customer trust, which is particularly crucial in specialized industries like flight and ground schools. According to a study by the Chartered Institute of Marketing, 57% of consumers do not trust brands to use their data responsibly. This lack of trust can be detrimental in building long-term customer relationships. Trust can be your most valuable asset or your biggest liability; once it's lost, it's tough to regain.

Legal compliance in email marketing is also a form of risk management. Lawsuits related to data breaches or misuse of customer information can impact your business financially and have a cascading effect on your overall brand value. A Ponemon Institute study showed that the average cost of a data breach in the U.S. is $8.64 million. This doesn't quantify the "soft costs" like brand erosion and lost consumer confidence, which can linger for years.

Being compliant also paves the way for better email deliverability. Emails that comply with legal standards are less likely to be flagged as spam, ensuring that your carefully crafted messages reach your audience's inbox. Therefore, it is imperative to understand and adhere to email marketing laws as they serve as a foundational element in the creation and execution of your digital marketing strategy, impacting everything from customer trust and brand reputation to financial stability.

CHOOSING A PROVIDER

Selecting the right email marketing platform is essential for aviation training businesses aiming to engage with their audience effectively. There are several popular email marketing platforms, each with its unique features and capabilities. Here, we'll compare a few of the prominent options to help you make an informed choice based on your flight school's specific needs:

1. Mailchimp

- **User-Friendly Interface:** Mailchimp is known for its drag-and-drop email builder, making it accessible to beginners.
- **Automation:** It offers robust automation features for drip campaigns, welcome sequences, and more.
- **Segmentation:** Allows you to segment your email list based on various criteria, allowing for highly targeted campaigns.
- **Analytics:** Provides detailed analytics and reports to track email performance.
- **Pricing:** Offers a free plan with limited features and paid plans based on the number of subscribers.

2. Constant Contact

- **Ease of Use:** Known for its ease of use and customizable templates.
- **Automation:** Offers automation tools to send personalized emails based on subscriber actions.
- **Event Marketing:** Integrates event marketing features for flight schools hosting seminars, webinars, or workshops.
- **Support:** Provides excellent customer support through phone, chat, and email.
- **Pricing:** Offers competitive pricing based on the number of contacts.

3. SendinBlue

- **Affordability:** Known for its competitive pricing, especially for businesses with large email lists.
- **Marketing Automation:** Provides advanced automation capabilities to nurture leads.
- **SMS Marketing:** Integrates SMS marketing, allowing you to reach subscribers via text messages.
- **Transactional Emails:** Offers transactional email capabilities for flight schools needing to send booking confirmations, receipts, etc.
- **A/B Testing:** Allows A/B testing to optimize email performance.

4. HubSpot

- **All-in-One Solution:** HubSpot offers a complete suite of marketing, sales, and CRM tools.
- **Email Personalization:** Allows advanced email personalization based on subscriber behavior.
- **Integration:** Seamlessly integrates with other HubSpot tools for comprehensive marketing automation.
- **Analytics:** Provides in-depth analytics for a 360-degree view of your email performance.
- **Pricing:** HubSpot offers a free CRM and email marketing tool, with paid plans based on the desired features.

5. ConvertKit

- **Creator-Focused:** Tailored for creators, bloggers, and educators, ConvertKit emphasizes simplicity and personalization.
- **Automation:** Provides straightforward automation for sequence emails and drip campaigns.
- **Tagging:** Uses a tagging system for segmenting subscribers, simplifying personalization.
- **Landing Pages:** Offers customizable landing pages for lead generation.
- **Pricing:** Pricing is based on the number of subscribers, with a free plan available for up to 1,000 subscribers.

Choosing the right email marketing platform ultimately depends on your aviation school's unique needs, budget, and goals. Consider factors such as ease of use, automation capabilities, pricing, and scalability when making your decision. It's also wise to take advantage of free trials or starter plans to explore these platforms firsthand and determine which one best aligns with your flight school's marketing strategy.

EMAIL MARKETING CHECKLIST

- [ ] **Segmentation:** Divide your email list based on key attributes such as the stage of training or areas of interest.
- [ ] **Content Planning:** Develop a content calendar to keep your email marketing consistent and valuable.
- [ ] **Analytics:** Track key metrics like open rates, click-through rates, and conversion rates to gauge the success of your campaigns.
- [ ] **A/B Testing:** Regularly run A/B tests to optimize various elements of your emails.
- [ ] **Automation:** Implement automated email series for specific actions like welcoming new students or post-training follow-ups.
- [ ] **Legal Compliance:** Ensure you adhere to all legal requirements related to email marketing.

LEVERAGE YOUR EMAIL MARKETING

This chapter has provided an in-depth guide on leveraging email marketing for flight and ground schools. When executed strategically, this tool can serve as a potent channel for both acquisition and retention, contributing to both immediate and long-term business success.

9

DISCOVER THE POWER OF COMMUNITY, WHEN BUILDING YOUR SOCIAL MEDIA MARKETING

CHAPTER NINE

Social Media Marketing

BUILDING A COMMUNITY OF AVIATION AMBASSADORS

In an era where digital presence can make or break a brand, social media marketing is a crucial element in the overall marketing strategy of any business. For specialized sectors like flight and ground schools, leveraging social media platforms can be a game-changer for building brand awareness, student engagement, and enrollment growth.

According to a study by Buffer, companies that invest in social media advertising are seeing 2-3 times—the ROI of traditional advertising methods. Social media isn't just for industries like fashion or technology; it's also incredibly pertinent for niche sectors like flight training.

SOCIAL MEDIA MARKETING
BUILD YOUR COMMUNITY

This chapter will explain how flight and ground schools can harness the power of social media for boosting their ROI. From platform selection to legal compliance, we'll walk you through the intricacies and arm you with actionable insights for immediate implementation. By the end of this chapter, you'll be ready to take your social media efforts to the next altitude.

PLATFORM SELECTION

Choosing the right platforms for your school is a critical first step. Facebook, Instagram, LinkedIn, and Twitter each offer unique advantages and reach different demographic groups. According to Pew Research, Instagram, is popular among consumers 18-29 years old, while LinkedIn is more frequented by professionals aged 25-34. Tailoring your social media marketing strategies to these platforms can yield better results and improved ROI.

LINKEDIN IS **277%** MORE EFFECTIVE FOR LEAD GENERATION THAN FACEBOOK AND TWITTER.

A study by Forrester Research revealed that LinkedIn is 277% more effective for lead generation than Facebook and Twitter, making it an excellent platform for B2B customer acquisition. With this in mind, you could start to think about segmenting your social media campaigns based on the type of audience you aim to attract. For instance, you can focus your LinkedIn content on decision makers for corporate contracts, while targeting younger aspiring pilots on Instagram.

Understanding platform mechanics is also essential. Each has its own set of algorithms that determine how content is displayed to users. For example, Instagram values engagement and relevancy, so posts that get immediate engagement in the form of likes, comments, and shares are more likely to be displayed prominently.

CONTENT STRATEGY

Content is the anchor that grounds your social media strategy. From informational videos and engaging stories to insightful articles, the type of content you share can significantly influence your audience. According to the Content Marketing Institute, 65% of the most successful organizations have a documented content strategy. A study by Adobe confirmed that posts with images produce a 650% higher engagement rate than text only posts. This emphasizes the importance of visually compelling content, particularly in an industry where showing the excitement and precision of flight can be a unique selling proposition.

But content isn't just about what looks good; it needs to resonate with your audience and represent your brand's voice and mission. For flight schools, this could mean going beyond showing sleek planes and picturesque views from the cockpit. Content that educates and informs, such as safety guidelines or breakdowns of different flight training courses, can also be invaluable. It provides prospective students with the information they need, positioning your school as a thought leader in the aviation industry.

It's also essential to have a content calendar that outlines what, when, and where to post. Timing can be critical in social media marketing. According to Sprout Social, the best times to post on Instagram are Wednesday at 11 a.m. and Friday between 10–11 a.m. For LinkedIn, the best times are Wednesday from 3–5 p.m. Knowing this, you can schedule your social content to go live during these high-traffic times to maximize engagement and, by extension, ROI.

ANALYTICS AND PERFORMANCE METRICS

Monitoring and analyzing your social media activities is essential for ROI optimization. Key performance indicators (KPIs) such as engagement rate, reach, and conversion can give invaluable insights. According to Sprout Social, brands that measure ROI are 1.6 times more likely to have higher budgets, emphasizing the necessity of a metrics-driven approach.

Not all metrics are created equal, though. While likes and shares are good for gauging general interest, they don't necessarily translate to enrollments or revenue. What you want to track are 'bottom-of-the-funnel' metrics, like click-through to your enrollment page or actual sign-ups triggered by a social media post.

Sophisticated tools like Google Analytics can track a user's journey from social media to enrollment. By using UTM parameters, you can trace the source of your website traffic back to individual social media campaigns, thereby measuring their effectiveness more accurately. Suppose you find that a specific campaign on LinkedIn is driving significant traffic but not conversions. That's a signal to revisit the landing page or offer associated with that campaign.

A SOCIAL MEDIA CONTENT CALENDAR IS ESSENTIAL FOR MARKETING SUCCESS.

PAID ADVERTISING AND ROI

While organic reach is vital, paid social media can provide immediate results and highly targeted traffic. A study by Nielsen revealed paid social ads can generate an 8% median ROI, offering both affordability and effectiveness. Knowing your Customer Lifetime Value (CLV) can further help you determine how much to invest in paid campaigns. By comparing the CLV to the Customer Acquisition Cost (CAC) through paid advertising, you can better optimize your ROI.

The versatility of paid advertising options across platforms also allows for highly segmented targeting, from geographic location to specific interests or behaviors. For flight schools, this could mean targeting ads specifically at people who have shown interest in aviation or educational courses, thereby increasing the likelihood of conversions.

WEBSITE VISITORS WHO ARE RETARGETED WITH DISPLAY ADS ARE **MORE LIKELY TO CONVERT BY 43%**

Also, retargeting campaigns can significantly enhance your paid advertising performance. If someone has visited your enrollment page but didn't complete the sign-up process, retargeting can serve them ads reminding them to complete their application. According to Criteo, website visitors who are retargeted with display ads are 43% more likely to convert.

LEGAL COMPLIANCE

Just like email marketing, social media also has legal guidelines. Failing to adhere can result in penalties and hurt your brand reputation. Adherence to Federal Trade Commission (FTC) regulations is a must. According to a report by the law firm Perkins Coie, 76% of businesses surveyed experienced social media-related legal risks, demonstrating that compliance isn't just a 'nice to have' but a necessity.

It's crucial to be transparent about your advertising methods. For example, if you're running an Influencer marketing campaign promoting your flight school, the FTC requires that these posts are clearly marked as sponsored content. This isn't just to avoid legal trouble; it's also about building and maintaining trust with your audience.

Data privacy is another significant concern, especially with stringent laws like the General Data Protection Regulation (GDPR) in the European Union. You're required to inform users how your company will use and story their data. Failure to comply can result in hefty fines that can severely impact your ROI.

CALENDAR

Creating a social media marketing calendar is a crucial step in ensuring a consistent and effective online presence for your aviation training business. Here's a concise guide on how to create one:

1. Set Clear Goals and Objectives

Begin by defining your social media marketing goals. These could include increasing brand awareness, driving website traffic, or generating leads for your flight school.

2. Identify Your Target Audience

Understand your audience's demographics, interests, and pain points. This information will help you tailor your content to resonate with them.

3.Choose the Right Platforms

Determine which social media platforms are most relevant to your audience and industry. Focus your efforts where your potential students are most active.

4. Content Planning

Plan your content themes and topics. Consider a balance of educational content, aviation insights, student success stories, and promotional posts for courses and events.

5. Content Calendar

Create a calendar that outlines what content will be posted on which days. Include dates, times, and specific content ideas.

6. Consistency is Key

Stick to a consistent posting schedule. Regular, predictable updates help build audience trust and engagement.

7. Visual Assets

Plan for visuals like images, videos, and graphics. High-quality, engaging visuals are essential for social media success.

8. Hashtags

Research relevant hashtags for your industry and incorporate them into your posts to increase discoverability. Also, create a branded hashtag that's unique to your school to include in your social content.

9. Engagement Strategy

Outline how and how often you'll engage with your audience. Respond to comments, messages, and participate in relevant discussions.

10. Analytics and Adjustments

Use social media analytics tools to track your performance. Analyze what's working and what needs adjustment, then refine your calendar accordingly.

11. Ad Campaigns

If you plan to run paid advertising campaigns on social media, incorporate these into your calendar with specific dates and budgets.

12. Holidays and Special Events

Include important holidays and aviation-related events in your calendar. Tailor your content to align with these occasions.

13. Team Collaboration

If you have a team, ensure everyone is on the same page regarding the content calendar and responsibilities.

14. Review and Adapt

Regularly review your social media calendar to assess its effectiveness. Adapt to changing trends, audience preferences, and business goals.

By following these steps and maintaining a well-organized social media marketing calendar, you can consistently deliver relevant and engaging content to your audience, ultimately strengthening your flight schools's online presence and achieving your marketing objectives.

CONCLUSION

Social media marketing is not a one-size-fits-all endeavor, particularly for specialized fields like flight and ground schools. By making informed choices about platforms, employing a robust content strategy, keeping an eye on analytics, considering paid advertising, and staying legally compliant, you can create a powerful social media marketing plan designed for sustainable ROI and growth.

Remember, achieving a high ROI in social media marketing doesn't just happen overnight. It takes consistent effort, constant learning, and ongoing adaptation to the ever-changing digital landscape. With a well-executed strategy, flight and ground schools can significantly enhance their ROI and set a trajectory for long-term success.

SOCIAL MEDIA OVERVIEW

	YOUTUBE	INSTAGRAM	TIKTOK	X (FORMERLY TWITTER)	FACEBOOK	LINKEDIN
ROLE	Educate & Enhance SEO	Entertain & Engage	Entertain & Inspire	Update & Educate	Entertain & Drive Leads	Credentialize
KEY DEMOGRAPHIC	15 - 35 51% M 49% F	18 - 24 52% M 48% F	18 - 24 54% M 46% F	18 - 29 66% M 34% F	25 - 34 56% M 44% F	30 - 39 57% M 43% F
AVG. DAILY USAGE	48.7 min.	33.1 min.	53.8 min.	34.1 min.	30.9 min.	7.27 min.
TYPES OF CONTENT	Course Overviews & Testimonials	Inspirational Videos & Images	Inspirational Videos	Topical Comments & Updates	Updates & Links	Thought Leadership & Company News
CONTENT LENGTH	2 min.	Stories - :15 Reels - :15 - : 60 Posts - 125 chars	7-15 sec.	71-100 chars	40-80 chars	100-140 chars
POSTING FREQUENCY	1x/week	1-2x/day	1-4x/day	3-4x/day	4-5x/day	1x/day

SOCIAL MEDIA PLANNING CALENDAR TEMPLATE

	MONDAY	TUESDAY	WEDNESDAY	THURSDAY	FRIDAY	SATURDAY	SUNDAY
LINKEDIN							
INSTAGRAM							
FACEBOOK							
TWITTER / X							
YOUTUBE							

SOCIAL MEDIA MARKETING CHECKLIST

- ☐ Select appropriate platforms based on your target audience.
- ☐ Develop a documented content strategy.
- ☐ Track KPIs regularly for actionable insights.
- ☐ Utilize both organic and paid social media strategies.
- ☐ Adhere to legal guidelines specific to social media marketing.

PLAN YOUR SOCIAL MEDIA MARKETING

By following these guidelines, flight and ground schools can significantly enhance their ROI and set a trajectory for sustained growth.

PART III

ADVANCED STRATEGIES

10

BALANCE YOUR BUDGET AND YOUR REACH WITH PAY-PER-CLICK ADVERTISING

CHAPTER TEN

Pay-Per-Click (PPC) Advertising

GETTING IMMEDIATE LIFT

Pay-Per-Click (PPC) advertising is an essential marketing tool, offering the immediate impact that many other methods can't provide. In the competitive field of aviation education, waiting for organic growth could mean watching potential students take flight with another school. According to Google, businesses make an average of $2 in revenue for every $1 spent on AdWords. This statistic serves as a compelling reason for flight and ground schools to leverage PPC advertising for quick, targeted results. This chapter will provide a comprehensive guide to harnessing the power of PPC advertising to propel your aviation training institute's growth.

PAY-PER-CLICK ADVERTISING

PPC PLATFORMS

Not all PPC platforms are created equal, especially when you're targeting a niche market like aviation training. Each platform has its own set of advantages and challenges. For instance, Google Ads offers the broadest reach, but it is also the most competitive. On the other hand, Bing Ads generally costs less on a per-click basis but reaches a more limited audience.

Choosing the right platform requires an understanding of where your prospective students are active online. A WordStream study reports the average conversion rate in Google Ads across all industries is 3.75%, but this varies widely between sectors. Knowing your audience's preferences, browsing habits, and needs can inform your choice of platform and consequently improve ROI.

THE AVERAGE CONVERSION RATE IN GOOGLE ADS ACROSS ALL INDUSTRIES IS **3.75%**

Selecting the appropriate PPC platform for your aviation training business is a pivotal decision that can significantly impact the success of your digital ad campaigns. Here's an exploration of some key platforms and considerations to help you make an informed choice:

1. Google Ads

Google Ads is one of the most widely used and powerful PPC platforms globally. It offers the broadest reach, with access to a vast user base actively searching for aviation-related terms. However, this broad reach also means heightened competition, which can result in higher costs per click (CPC).

Google Ads provides various ad formats, including text ads, display ads, and video ads, allowing you to tailor your strategy to your aviation training offerings. Moreover, its robust keyword targeting capabilities enable precise ad placement.

2. Bing Ads

While much less prominent than Google, Bing offers its own advantages. It tends to have lower CPCs compared to Google Ads, making it a cost-effective choice for aviation training businesses aiming to maximize their budget. Although it reaches a smaller audience than Google, Bing still has millions of users actively searching for information.

3. Social Media Platforms

Social media platforms like Facebook and LinkedIn also offer PPC advertising options. These platforms can be particularly valuable if your aviation training business has a strong presence on social media or if you're looking to target specific demographics or job titles.

Choosing the Right Platform

To select the right PPC platform for your flight school, start by understanding where your prospective students spend their time online. Are they primarily using search engines to research flight training options, or do they frequent social media platforms for aviation-related content and networking?

Additionally, consider your budget and the level of competition within your chosen platform. While Google may offer broader reach, it can be more competitive and expensive. Conversely, Bing Ads can provide cost savings, but with a potentially smaller audience.

Understanding your audience's preferences, browsing habits, and needs is crucial in making an informed decision. Conduct thorough market research and consider running small-scale test campaigns on different platforms to gauge their effectiveness in reaching your aviation training prospects. Monitoring the results of these campaigns can provide valuable insights into which platform offers the best ROI for your flight school.

In conclusion, selecting the right PPC platform for your aviation training business should be based on a comprehensive understanding of your target audience and the platform's suitability for your goals and budget. By aligning your choice with your audience's online behavior, you can optimize your PPC campaigns and maximize their impact on driving leads and conversions for your flight school.

CHOOSING THE RIGHT PPC PLATFORM REQUIRES KNOWING YOUR BUDGET AND KNOWING WHERE YOUR PROSPECTIVE STUDENTS SPEND THEIR TIME ONLINE.

Here's a brief summary comparing when to use these platforms:

1. Google Ads

Benefits: Offers the broadest reach with a vast user base actively searching for aviation-related terms. Robust keyword targeting capabilities. Various ad formats available, including text, display, and video ads.

Drawbacks: High competition can result in higher costs per click (CPC). Advertisers with limited budgets may find it challenging to compete effectively.

2. Bing Ads

Benefits: Generally offers lower CPCs compared to Google Ads, making it cost-effective. Still reaches a sizable audience actively searching for information.

Drawbacks: Has a smaller user base compared to Google, limiting reach. May not be as suitable for businesses with a strong focus on Google Ads.

3. Social Media Platforms (e.g., Facebook, LinkedIn)

Benefits: Valuable for targeting specific demographics or job titles. Suitable for aviation training businesses with a strong social media presence.

Drawbacks: May not have as direct or extensive search intent as search engine platforms like Google and Bing. Ad effectiveness can vary based on the platform and audience targeting.

Choosing the Right Platform

Use Google Ads when you seek a broad reach and have a budget to compete in a competitive space. Ideal for aviation training businesses with diverse ad format needs.

Consider Bing Ads when budget optimization is a priority, and you still want to reach a substantial audience. Suitable for cost-conscious aviation training businesses.

Use Social Media Platforms when targeting specific demographics, such as aspiring pilots or aviation professionals. Ideal for aviation schools with a strong social media presence and content strategy.

Selecting the right PPC platform should be driven by your understanding of your target audience's online behavior, your budget, and the level of competition in your chosen platform. Each platform offers unique advantages and drawbacks, so align your choice with your aviation training business's specific goals and resources to optimize your PPC campaigns effectively.

KEYWORD RESEARCH

The power of PPC advertising lies in its ability to target specific queries entered by users. Keyword research, therefore, becomes the cornerstone of any successful PPC campaign.

Understanding user intent is also essential. Keywords can be broadly classified into informational, navigational, and transactional. Informational keywords could be phrases like "how to become a pilot," while transactional keywords could be "CPL course fees." These nuances can profoundly impact the cost and effectiveness of your PPC campaigns.

Unlocking the Power of Keywords: A Step-by-Step Guide for Flight Schools

In advertising, the ability to precisely target user queries through keywords is a superpower. This step-by-step guide will walk you through the process of choosing the right keywords, understanding user intent, and comparing keyword performance effectively.

70% OF THE SEARCH SHARE IS MADE UP OF LONG-TAIL KEYWORDS.

Step 1: Start with Seed Keywords

Begin your keyword research journey by identifying a list of seed keywords related to aviation training. These are broad terms that encapsulate the essence of your flight school's offerings. Examples include "flight school," "aviation training," or "pilot courses." These seed keywords will create the foundation for your research.

Step 2: Expand with Long-Tail Keywords

Dive deeper into keyword research by expanding from your seed keywords to long-tail keywords. Long-tail keywords are more specific and often consist of three or more words. For aviation training, these could include phrases like "CPL training in [Your Location]," "Private pilot ground school near me," or "Instrument rating course cost."

Long-tail keywords are crucial for flight schools because they capture the specificity of user intent. As the statistics reveal, approximately 95% of search share is comprised of long-tail keywords.[19] Users who search for specific phrases are often closer to making a decision, so these keywords are highly valuable for PPC campaigns.

Step 3: Understand User Intent

Understanding user intent is key to selecting the right keywords for your flight schol and prospective students. Keywords can be broadly categorized into three types based on intent:

1. Informational Keywords

These keywords are used by users seeking information or answers to their queries. Examples include "how to become a pilot," "aviation career paths," or "pilot training requirements." Informational keywords are valuable for educational content and blog posts.

2. Navigational Keywords

Navigational keywords are used by users looking for specific websites or resources. Examples include "ABC Flight School website" or "online pilot courses." While not typically used for PPC campaigns, they are essential for optimizing your flight school's website.

3. Transactional Keywords

These keywords indicate a user's intent to take action, such as enrolling in a course or making a purchase. Examples include "CPL course fees," "flight school enrollment," or "aviation training discounts." Transactional keywords are the heart of your PPC campaign, as they directly drive conversions.

Step 4: Use Keyword Research Tools

Leverage keyword research tools to refine your list of keywords. Tools like Google's Keyword Planner, SEMrush, or Ahrefs can provide valuable insights into search volume, competition, and related keywords.

For instance, let's compare two keywords: "Private pilot training" and "Private pilot ground school near me." Using a keyword research tool, you discover that "Private pilot training" has a high search volume and high competition. On the other hand, "Private pilot ground school near me" has a lower search volume and significantly lower competition.

In this scenario, you may choose to prioritize "Private pilot ground school near me" for your PPC campaign because it aligns with users in your specific location who are likely seeking immediate enrollment in ground school. This targeted approach can result in higher conversion rates and a more cost-effective campaign.

Step 5: Assess Keyword Performance

Once your PPC campaign is up and running, it's essential to continuously assess keyword performance. Most PPC platforms provide detailed analytics to help you understand which keywords are driving results.

Compare the performance of different keywords by examining metrics such as:

- **Click-Through Rate (CTR):** The percentage of users who clicked on your ad after seeing it. A high CTR indicates that a keyword is relevant and enticing to your audience.
- **Conversion Rate:** The percentage of users who took a desired action after clicking on your ad, such as filling out a contact form or enrolling in a course. A higher conversion rate suggests that a keyword is effective in driving conversions.
- **Cost Per Click (CPC):** Analyze the cost associated with each click for a particular keyword. Keywords with a lower CPC may offer a better return on investment.
- **Return on Investment (ROI):** Calculate the overall ROI for each keyword by considering the cost of advertising and the revenue generated. High-ROI keywords should be prioritized in your campaign.

In Conclusion

Keyword research is an ongoing and iterative process. By starting with seed keywords, expanding to long-tail keywords, understanding user intent, using keyword research tools, and assessing keyword performance, you can optimize your PPC campaign effectively. The ability to choose the right keywords and adapt based on real-time data will help you determine the most cost-effective way for your aviation training businesses to attract and convert prospective students.

AD COPY AND LANDING PAGES

Once you've zeroed in on your keywords, the next crucial step is crafting compelling ad copy and optimizing your landing pages. According to Unbounce, the average conversion rate for PPC landing pages is around 9.7%. However, optimized landing pages can reach up to 27.4% conversion rates, showcasing the immense potential in refining this element of your PPC strategy. This topic is so important we've dedicated an entire chapter (Chapter 10: "Ad Copy and Landing Pages") to it.

BIDDING STRATEGY

One of the more complex aspects of PPC advertising is deciding on your bidding strategy. Choices range from automated strategies like Maximize Clicks or Target CPA to manual strategies where you set bids for each keyword. 97% of Google Ads advertisers used automated bidding strategies in 2020, per SEMrush. However, manual bidding can offer more control over specific campaign elements, allowing for better alignment with your ROI goals.

Different strategies work better for different campaign objectives. For example, if you're looking to quickly increase enrollments for a new course, you might opt for a Target ROAS (Return On Advertising Spend) strategy to maximize revenue for a set budget. Alternatively, if brand visibility is the primary objective, a Maximize Clicks strategy might be more suitable.

MANUAL BIDDING CAN OFFER **MORE CONTROL OVER SPECIFIC CAMPAIGN ELEMENTS**, ALLOWING FOR BETTER ALIGNMENT WITH YOUR ROI GOALS.

Modifying Bidding Strategy Based on ROAS (Return On Advertising Spend)

In the intricate world of PPC advertising, choosing the right bidding strategy is a critical component of campaign success. The approach you select can significantly impact your aviation training business's ROI. Let's delve into how a flight school marketer would modify their bidding strategy based on ROAS.

1. Understanding Bidding Strategies

There are two primary categories of bidding strategies in PPC advertising: automated and manual.

- **Automated Bidding:** These strategies, such as Maximize Clicks or Target CPA (Cost Per Acquisition), rely on machine learning algorithms to optimize bids automatically. They are favored for their convenience and efficiency, particularly by advertisers with large campaigns.

- **Manual Bidding:** In contrast, manual bidding allows advertisers to set specific bids for each keyword or ad group. While it demands more hands-on management, it offers greater control over individual elements of the campaign.

2. The Role of ROAS

Return On Advertising Spend (ROAS) is a metric that quantifies the revenue generated from your ad spend. It's a key indicator of campaign profitability and is instrumental in determining how efficiently your advertising dollars are being utilized.

3. Modifying Bidding Strategy with ROAS

Choosing the right bidding strategy depends on your campaign objectives and how you prioritize ROAS in your aviation training business.

CHOOSING THE RIGHT BIDDING STRATEGY IS **A CRITICAL COMPONENT** OF CAMPAIGN SUCCESS.

a. Target ROAS Strategy

The Target ROAS bidding strategy is most suitable for flight schools looking to maximize revenue while maintaining a set budget. Here's how it works:

- Set a specific ROAS target, such as 300%, meaning you aim to generate $3 in revenue for every $1 spent on advertising.
- Google's automated system adjusts your bids in real-time to achieve the desired ROAS.
- This strategy is ideal when your primary goal is to increase enrollments for a new course or generate substantial revenue from your ad spend.

b. Maximize Clicks Strategy

Conversely, the Maximize Clicks bidding strategy prioritizes driving as much traffic as possible within your budget. While it may not directly optimize for ROAS, it can be valuable for other campaign objectives:

- If brand visibility is your primary goal, Maximize Clicks can help attract a broader audience to your flight school's website.
- You can still monitor ROAS and make adjustments as needed to align with profitability goals, but the primary focus here is on increasing website traffic.

c. Manual Bidding for Precision

While automated bidding strategies are popular, manual bidding offers distinct advantages, especially when your goal is precise control over keyword-level bids and ROAS:

- By manually setting bids for specific keywords or ad groups, you can allocate budget strategically to maximize ROAS for high-converting keywords.
- This approach provides flexibility to adjust bids based on historical data and emerging trends, ensuring you stay in line with your ROAS objectives.

4. Continuous Monitoring and Optimization

Regardless of the bidding strategy chosen, ongoing monitoring and optimization are essential. Flight school marketers should regularly assess ROAS performance, adjust bids based on keyword ROI, and refine the strategy to maximize profitability.

In conclusion, modifying your bidding strategy based on ROAS is a dynamic process. The choice between automated and manual strategies depends on your campaign goals and the level of control you desire. The Target ROAS strategy is a valuable tool for maximizing revenue while maintaining efficiency, while the Maximize Clicks strategy can enhance brand visibility. Manual bidding offers precision control for optimizing ROAS at the keyword level. The key is to align your bidding strategy with your aviation training business's specific objectives and continually refine it to achieve the desired ROAS and campaign success.

ANALYTICS AND ROI

Monitoring analytics is a critical success factor of your PPC campaigns. Key metrics like Click-Through Rate (CTR), Cost Per Conversion, and Quality Score need constant tracking. According to a report by HubSpot, companies that track metrics are 17% more likely to achieve their goals.

Besides, understanding metrics like Lifetime Value (LTV) of a customer against the Cost Per Acquisition (CPA) can provide deep insights into your PPC campaign's ROI.

Advanced analytics platforms can offer a treasure trove of data, including the customer journey from click to conversion. Google Analytics, for instance, allows you to set up conversion tracking to monitor how many clicks on your ads resulted in enrollments, thereby offering direct insight into ROI.

COMPANIES THAT TRACK METRICS ARE **17%** MORE LIKELY TO ACHIEVE THEIR GOALS.

LEGAL COMPLIANCE

Similar to social media marketing, PPC advertising comes with its own set of legal responsibilities. Compliance with Federal Trade Commission (FTC) guidelines, as well as international regulations like GDPR for European customers, is essential. Non-compliance can result in hefty fines and severely tarnish your reputation, negating any positive ROI from your PPC campaigns.

Transparency is critical, especially when dealing with sensitive industries like aviation. For instance, if your ad claims a 100% job placement rate after graduation, you must have data to back that up. Inaccurate or misleading claims can lead to regulatory action and consumer mistrust.

CONCLUSION

PPC advertising provides flight and ground schools with the opportunity for immediate growth and improved ROI. By choosing the right platforms, conducting thorough keyword research, optimizing ad copy and landing pages, setting a smart bidding strategy, and diligently monitoring analytics, you can ensure that your PPC campaigns soar. Legal compliance remains a non-negotiable aspect, keeping you clear of turbulence that could derail your PPC success.

PAY-PER-CLICK ADVERTISING CHECKLIST

- ☐ Choose the most appropriate PPC platform based on your target audience.
- ☐ Conduct in-depth keyword research, considering the nuances of user intent.
- ☐ Craft compelling ad copy and optimize landing pages for conversions.
- ☐ Decide on a bidding strategy aligned with your campaign objectives.
- ☐ Monitor key metrics regularly for actionable insights.
- ☐ Adhere to legal guidelines specific to PPC advertising.

DRIVE YOUR PAY-PER-CLICK ADVERTISING

By adhering to these guidelines, flight and ground schools can significantly elevate their ROI and secure a pathway for immediate and sustained growth.

11

LEARN HOW TO CREATE COPY THAT ENGAGES YOUR AUDIENCE AND DRIVES ENGAGEMENT

CHAPTER ELEVEN

Ad Copy and Landing Pages

CONVERTING INTEREST INTO ACTION

Once you've zeroed in on your keywords, the next crucial step is crafting compelling ad copy and optimizing your landing pages. Per Unbounce, the average conversion rate for PPC landing pages is around 9.7%. However, optimized landing pages can reach up to 27.4% conversion rates, showcasing the immense potential in refining this element of your PPC strategy.

Your ad copy should be enticing, but it must also be aligned with the content on your landing page. Inconsistency between these can not only lead to lower conversion rates but can also result in higher costs per click due to poor Quality Scores. Your ad's Quality Score, a metric used by Google Ads, takes into account the relevance of your ad copy, keywords, and landing page, impacting both ad placement and cost.

AD COPY

Choosing effective ad copy for your aviation training business requires a strategic approach. Consider the following formula to craft compelling ad copy that resonates with your target audience and encourages action:

1. Identify Key Benefits

Highlight the most significant benefits your flight school offers. What sets you apart from the competition? Whether it's experienced instructors, state-of-the-art simulators, or a high graduation rate, identify your unique selling points (USPs).

2. Understand User Intent

Tailor your ad copy to match the user's intent. Are they looking for private pilot training, instrument rating courses, or career advice? Ensure your ad copy aligns with their specific needs.

3. Use Action-Oriented Language

Employ action-oriented language to prompt users to take action. Phrases like "Enroll Now," "Start Your Journey," or "Get Certified Today" convey a sense of urgency and drive users to click.

4. Include Keywords

Integrate relevant keywords into your ad copy to improve ad relevance and quality score. Keywords should flow naturally within the ad text.

5. Address Pain Points

Identify and address common pain points or doubts that potential students may have. For example, if cost is a concern, mention any financial aid options or scholarships available.

6. Highlight Social Proof

Incorporate student success stories, testimonials, or any awards or certifications your flight school has received to build trust and credibility.

7. Create a Sense of FOMO

Utilize the fear of missing out by including phrases like "Limited Spots Available" or "Join Hundreds of Successful Graduates."

8. A/B Testing

Don't rely on a single ad. Conduct A/B testing to compare different ad variations and determine which ones perform best.

EXAMPLES OF SUCCESSFUL ADVERTISING COPY

Images: https://www.wordstream.com/blog/ws/2021/09/21/advertising-slogans, https://uberbuttons.com/blog/50-best-slogans-taglines/

LANDING PAGE EXAMPLE

Following is a template for a landing page that effectively guides prospective students from interest to action while addressing their doubts:

AVIATION TRAINING
YOUR JOURNEY BEGINS HERE

WELCOME TO [YOUR FLIGHT SCHOOL]

At **[Your Flight School]**, we understand that embarking on a journey in aviation can be both exciting and filled with questions. Whether you're dreaming of becoming a pilot or seeking advanced aviation training, our comprehensive programs and experienced instructors are here to guide you every step of the way.

WHY CHOOSE [YOUR FLIGHT SCHOOL]?

1. Unparalleled Expertise

With decades of experience in aviation education, we're proud to be leaders in the industry. Our instructors aren't just teachers; they're mentors who are passionate about helping you succeed.

2. Cutting-Edge Facilities

Train in state-of-the-art facilities equipped with the latest technology and resources. Our flight simulators provide a realistic experience that sets you up for success.

3. Proven Track Record

Join the ranks of our successful graduates who have forged ahead to fulfilling careers in aviation. Our high graduation rate speaks volumes about the quality of education we provide.

ADDRESSING YOUR CONCERNS

We know you might have doubts or questions, and that's perfectly normal. Here's how we address some common concerns:

1. **Cost and Financing**

   We offer flexible payment plans, financial aid options, and scholarships to make your dream of flying more accessible.

2. **Safety First**

   Your safety is our top priority. Our rigorous safety protocols and commitment to excellence ensure you're in good hands.

3. **Flexible Scheduling**

   We understand life can be busy. That's why we offer flexible scheduling options to accommodate your needs.

JOIN OUR AVIATION COMMUNITY TODAY

Ready to take the first step? Click below to schedule a consultation with our admissions team or to enroll in one of our programs. We're here to answer all your questions and provide the support you need to embark on your aviation journey with confidence.

SCHEDULE A CONSULTATION

ENROLL NOW

Still have questions? Feel free to contact us at [Contact Information]. We're here to help you make an informed decision about your aviation training.

AT [YOUR FLIGHT SCHOOL], YOUR DREAMS TAKE FLIGHT!

This landing page effectively addresses the key benefits of your flight school, answers common doubts, and guides prospective students towards taking action. It provides clear calls to action that prompt users to schedule a consultation or enroll, and it reinforces your flight school's credibility and commitment to student success.

EXAMPLES OF SUCCESSFUL INFOGRAPHICS

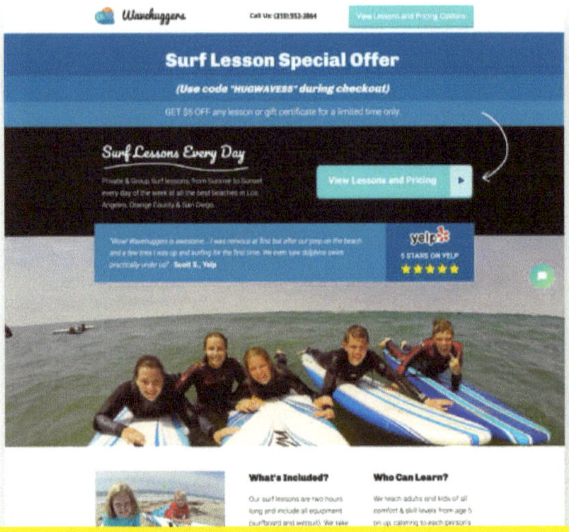

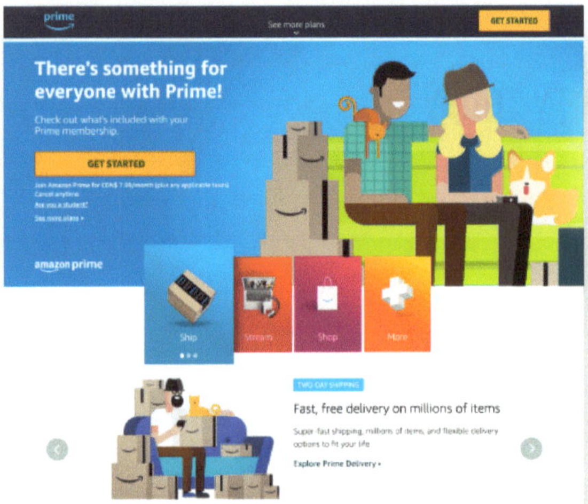

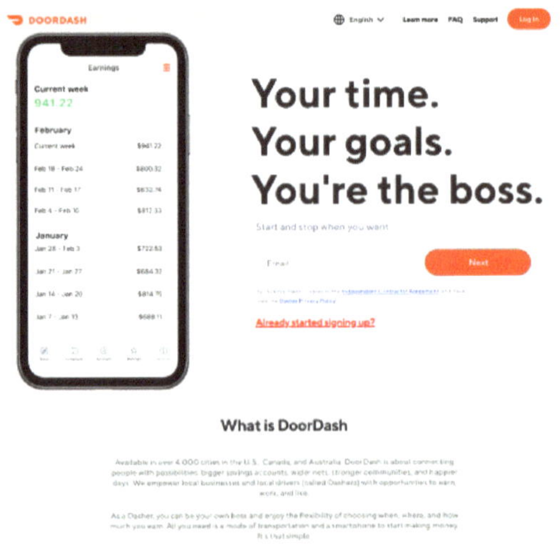

Images:
https://unbounce.com/landing-page-examples/best-landing-page-examples/

MASTER YOUR LANDING PAGE

By following these principles for building a successful landing page, flight and ground schools can leverage their credibility and drive growth and new business.

12

DISCOVER HOW TO INCREASE YOUR TRAFFIC WITH THE POWER OF AFFILIATE MARKETING

CHAPTER TWELVE

Affiliate Marketing

GENERATE LEADS THROUGH THIRD-PARTY PUBLISHERS

Affiliate marketing is often overlooked in education sectors like flight and ground schools, yet its potential for generating consistent, high ROI is significant. According to OptInMonster, affiliate marketing is worth more than $17 billion as of 2023, underscoring its increasing effectiveness as a digital marketing strategy.[20] When done correctly, affiliate marketing can create a seamless, symbiotic relationship between your flight school and affiliates that leads to mutual growth. This chapter will explore the facets of affiliate marketing and demonstrate how it can be strategically implemented to boost both ROI and overall growth.

AFFILIATE MARKETING

POWER THROUGH PARTNERSHIPS

IDENTIFYING AFFILIATES

JUST AS PILOTS MUST **CHOOSE THEIR CO-PILOTS WISELY,** SO MUST FLIGHT SCHOOLS IN THE REALM OF AFFILIATE MARKETING.

Affiliate marketing is an agreement to pay a commission to an individual or partner brand to promote your flight school through a website, blog, email or social media. Just as pilots must choose their co-pilots wisely, you must be discerning in selecting affiliate marketing partners. A poor choice can lead to brand misalignment or even reputational risk. On the other hand, a well-matched affiliate can drive highly-targeted traffic and potential students your way. Research by AM Navigator suggests 40% of marketers consider affiliate marketing's ability to acquire new customers as essential.

Potential affiliates might include flight training bloggers, aviation equipment stores, or even alumni who have built their own platforms in the aviation field. These affiliates should already be reaching your target demographic—people interested in aviation training. Once you've vetted them to confirm their industry credibility and they've agreed to partner with you, draft a straightforward but also detailed contract, outlining the responsibilities, payouts, and contingencies, thereby ensuring a transparent partnership beneficial to both parties.

COMMISSION STRUCTURES

The next consideration is how you'll compensate affiliates for driving students your way. According to a study by the Rakuten Marketing network, 94% of publishers prefer performance-based models like cost per action (CPA), where they get paid for every lead or sale they produce. While this can offer immediate ROI, there might be scenarios where a flat fee or recurring commissions could be more appropriate.

For example, if an affiliate's audience has a high lifetime value or is more likely to enroll in expensive courses, a revenue-sharing model might be more attractive. Flight schools should weigh the pros and cons of each approach, considering the long-term ROI and customer acquisition costs. Always remember that a greedy approach to commissions may discourage quality affiliates, thereby undermining the partnership and potential ROI.

TRACKING AND ATTRIBUTION

One of the most critical aspects of affiliate marketing is tracking. The success of your program will largely hinge on how accurately you can attribute leads, click-throughs, and registrations to individual affiliates. CMOs report that attribution is fully in place in 58% of companies, which allows for a more precise understanding of marketing ROI, according to Nielsen.

Various tracking methods, from UTM parameters to cookie tracking and advanced software solutions, can provide a high level of detail on user behavior, ensuring that affiliates are appropriately rewarded for their efforts. Precise tracking not only builds trust but also allows for better optimization. With comprehensive data, you can quickly identify which affiliates are delivering the most ROI, allowing for reallocation of resources and budget as needed.

AFFILIATE MARKETING AND SEO

While affiliate marketing's primary goal is direct revenue generation, it can also offer significant SEO benefits. Quality backlinks from high-traffic affiliates can improve your site's search engine rankings, increasing organic traffic over time. According to Moz, backlinks are one of the top factors for Google's algorithm for search rankings.

However, there's a caveat. Search engines have become increasingly adept at identifying "unnatural" backlinking strategies. Therefore, your affiliate partnerships should be genuine collaborations with quality content that brings value to the consumer, rather than a thinly-veiled excuse for backlinking. When properly executed, an affiliate marketing program can yield both immediate ROI through direct referrals and long-term ROI through improved search engine visibility.

> ONE OF THE **MOST CRITICAL ASPECTS** OF AFFILIATE MARKETING IS TRACKING.

LEGAL ASPECTS

You can't afford to overlook legal compliance when establishing your affiliate marketing program. According to the FTC, failure to disclose affiliate relationships can result in penalties. Ensure all affiliates disclose their connection to your flight school when they're promoting your services or courses. Non-compliance could not only result in legal repercussions but also damage your school's reputation, thereby negatively impacting ROI.

CONCLUSION

Affiliate marketing offers flight and ground schools a viable strategy for sustainable growth. By carefully selecting affiliates, crafting an attractive commission structure, implementing accurate tracking, considering SEO benefits, and adhering to legal guidelines, your affiliate program can become a significant revenue generator. The key is to view affiliates as long-term partners in growth rather than merely as channels for short-term revenue gains.

AFFILIATE MARKETING CHECKLIST

- [ ] Identify potential affiliates that align with your brand and target audience.
- [ ] Determine a commission structure that benefits both parties.
- [ ] Implement precise tracking methods for transparent attribution.
- [ ] Utilize affiliate marketing for dual benefits—direct revenue and SEO.
- [ ] Always adhere to legal guidelines and ensure affiliates do the same.

EMPOWER WITH AFFILIATE MARKETING

By comprehensively understanding and applying these elements, flight and ground schools can maximize their ROI and set the stage for long-term growth and success.

13

A WISE USE OF INFLUENCER MARKETING CAN INCREASE YOUR RELEVANCE IN THE INDUSTRY

CHAPTER THIRTEEN

Influencer Marketing

UTILIZING TRUSTED VOICES IN AVIATION

In the hyper-connected digital age, word-of-mouth has transcended local coffee shops and is now proliferating through social media platforms, blogs, and vlogs. Influencer marketing, the practice of partnering with influential individuals to promote your brand or service, has become a mainstay in the digital marketing toolkit. According to the Influencer Marketing Hub, the market size for influencer marketing is expected to grow to $21.1 billion in 2023.[21] For flight and ground schools, leveraging the power of influencers can be instrumental for driving growth and maximizing ROI. This chapter will examine how influencer marketing can be implemented effectively in the specialized niche of aviation training.

INFLUENCER MARKETING

THE RIGHT FIT

Just as you wouldn't pair a novice student with a complicated aircraft, your influencer partnerships should also be thoughtfully matched. A report by MediaKix states that, 89% of marketers find ROI from influencer marketing comparable to or better than other marketing channels. However, the effectiveness largely hinges on selecting the right influencers. Potential candidates may include pilots with significant followings, aviation enthusiasts, and even past successful students of your program.

When evaluating influencers, look beyond the follower count. Analyze their engagement rates, the quality of their audience, and how well their brand persona aligns with your flight school's ethos. The key is to find influencers who naturally fit with your brand and have a following that represents your target student demographic. Influencers with a dedicated, albeit smaller, following (often referred to as micro-influencers) can sometimes yield a higher ROI due to the quality and engagement level of their audience.

THE EFFECTIVENESS OF INFLUENCER MARKETING DEPENDS LARGELY ON **SELECTING THE RIGHT INFLUENCER.**

ADVANCED STRATEGIES FOR INFLUENCER MARKETING

1. Influencer Collaborations

Beyond one-off promotions, consider long-term partnerships that can foster a more authentic and sustained connection with your audience. For instance, you could work with an influencer to create a series of in-depth videos about different aspects of your training programs. This not only keeps your school in the spotlight but also provides valuable, ongoing content for your audience.

2. Co-Creation of Content

Instead of merely promoting your school, involve influencers in the co-creation of content. This could involve joint webinars, panel discussions, or even influencer-led courses within your school's offerings. Such collaborations not only expand your school's reach but also leverage the influencer's expertise to provide unique value to your students.

3. Influencer Takeovers

Allow influencers to take over your school's social media accounts for a day or a week. During this time, they can provide an insider's view of your school, showcasing training facilities, interviews with instructors, and interactions with students. Takeovers can boost engagement by giving your audience an authentic glimpse into what your school offers.

4. Contests and Challenges

Encourage influencers to launch aviation-related contests or challenges. This could involve creating flight-related content, such as landing challenges, navigation quizzes, or artistic interpretations of aviation. Contests can generate buzz, engage the audience, and even go viral, significantly amplifying your school's visibility.

EMERGING TRENDS IN INFLUENCER MARKETING

1. Nano-Influencers

While micro-influencers have been popular, nano-influencers (those with fewer than 1,000 followers) are gaining traction. They offer highly niche and engaged audiences, making them ideal for specialized industries like aviation training.

WHEN EVALUATING INFLUENCERS, **LOOK BEYOND** THE FOLLOWER COUNT.

2. Live Streaming

The rise of live streaming platforms presents new opportunities for influencer marketing. Influencers can host live Q&A sessions, virtual tours of your school, or conduct real-time flight simulations, providing immediate interaction with potential students.

3. Sustainability Influencers

As environmental concerns become more prominent, collaborating with influencers who focus on sustainability in aviation can position your school as forward-thinking and environmentally responsible.

4. Interactive Content

Interactive content like quizzes, polls, and challenges can enhance engagement and interactivity with your audience. Influencers can play a pivotal role in creating and promoting these interactive experiences.

5. AI-Powered Influencer Matching

Utilizing artificial intelligence to identify the best influencer matches for your school based on their content, audience, and engagement metrics will ensure more precise influencer selection.

CREATING AUTHENTIC CONTENT

Trust is an invaluable commodity in the aviation training industry, where safety and quality instruction are paramount. Influencers should therefore create authentic content that aligns with these core values. 86% of consumers say authenticity is essential when deciding which brands to support, per a study by Stackla.

Genuine experiences create authentic content, which resonates more with potential students. Involve influencers in real-life experiences at your flight school. Let them participate in a trial class, use your flight simulators, or interview students and instructors. The more relatable and less "salesy" the content is, the more likely it is to inspire action.

METRICS AND ROI

Like any other marketing strategy, influencer marketing needs to be measured to assess its impact accurately. Earned media value is the most commonly used metric, according to a report by Influencer Marketing Hub, in which 89% of marketers said they use it to measure their influencer marketing ROI.

Track metrics like engagement rate, click-throughs to your website, and conversion rates of leads into actual students. The data you collect can help you understand what's working, what needs fine-tuning, and how to allocate your budget most effectively for future campaigns. These metrics can also help you establish long-term partnerships with influencers who have proven their ability to deliver results.

Accurately measuring the ROI of influencer marketing is crucial for optimizing your strategies. Here are some KPIs and measurement methods to consider:

86% OF CONSUMERS SAY AUTHENTICITY IS ESSENTIAL WHEN DECIDING WHICH BRANDS TO SUPPORT.

1. Engagement Rate

Analyze the influencer's content engagement rate, including likes, comments, shares, and overall reach. High engagement indicates the effectiveness of the campaign in connecting with the audience.

2. Click-Through Rate (CTR)

Track the CTR on influencer-generated content leading to your school's website or landing pages. This metric measures the effectiveness of the content in driving traffic.

3. Conversion Rate

Measure the rate at which leads from influencer marketing campaigns convert into enrolled students. This is a critical metric for assessing the impact on actual ROI.

4. Cost-Per-Action (CPA)

Calculate the CPA by dividing the total campaign cost by the number of desired actions, such as enrollments. A lower CPA signifies more efficient spending.

TRUST IS AN INVALUABLE COMMODITY IN THE AVIATION TRAINING INDUSTRY.

5. Lifetime Value (LTV)

Determine the LTV of students acquired through influencer marketing. This metric assesses the long-term revenue generated by each student.

6. Brand Mentions and Sentiment Analysis

Monitor online mentions of your school generated by influencer campaigns. Assess the sentiment surrounding these mentions to gauge brand perception.

7. Attribution Modeling

Implement advanced attribution models to understand how influencer touchpoints contribute to the conversion journey. This provides insights into the influence of influencers on final enrollments.

LEGAL COMPLIANCE AND TRANSPARENCY

Similar to affiliate marketing, legal compliance is non-negotiable in influencer marketing. Federal Trade Commission (FTC) guidelines require that influencers disclose any paid partnerships or complimentary services clearly and conspicuously. According to Sprout Social, failing to disclose can result in fines up to $11,000 per violation.

Train your influencers on how to make disclosures appropriately, whether it's by using hashtags like #ad or #sponsored or making verbal acknowledgments in video content. This will not only keep you on the right side of the law but also add an extra layer of transparency and trust with your audience.

CONCLUSION

Influencer marketing can serve as a powerful vehicle for boosting both your ROI and overall brand reputation in the flight and ground school market. By judiciously selecting influencers, encouraging the creation of authentic content, vigilantly tracking performance metrics, and adhering to legal requirements, your school can yield significant benefits from this modern marketing strategy.

By effectively integrating influencer marketing into your broader digital strategy, you pave the way for more meaningful connections, greater student enrollments, and a substantially better ROI.

INFLUENCER MARKETING CHECKLIST

- [ ] Explore advanced influencer marketing strategies such as long-term collaborations and co-creation of content.
- [ ] Learn from real-world case studies to glean insights into the effectiveness of influencer marketing.
- [ ] Stay ahead of emerging trends, including the rise of nano-influencers and sustainability influencers.
- [ ] Measure the ROI of influencer marketing campaigns using KPIs like engagement rate, CTR, conversion rate, and CPA.
- [ ] Ensure legal compliance and transparency in influencer partnerships by adhering to FTC guidelines and crafting comprehensive contracts.

EMPOWER YOUR INFLUENCER MARKETING

By effectively integrating influencer marketing into your broader digital strategy, you pave the way for more meaningful connections, greater student enrollments, and a substantially higher ROI.

14

BUILD TRUST AND CREDIBILITY WITH POTENTIAL STUDENTS BY UTILIZING VIDEO MARKETING

CHAPTER FOURTEEN

Video Marketing

A VIRTUAL TOUR OF YOUR TRAINING FACILITIES

In today's digital landscape, video has become a dominant form of communication for brands. According to Social Pilot, more than 78% of viewers consume video content every week, and 55% of them engage every day.[22] The use of video in marketing provides flight and ground schools an excellent opportunity to showcase their training facilities, aircraft, instructors, and student life. While traditional means of marketing can describe the features and advantages of your aviation training program, video can actually show it. And let's not forget: "seeing is believing."

According to Wyzowl's Video Marketing Statistics 2022 report, 86% of businesses use video as a marketing tool, and a staggering 93% of marketers who use video say it's a critical part of their marketing strategy. Why? Because video content delivers results. It engages viewers, boosts conversion rates, and provides a higher ROI than other forms of content. For the niche sector of aviation training, video marketing offers an unmatched opportunity to virtually transport potential students into your state-of-the-art training facilities and high-tech cockpits.

VIDEO MARKETING

THE PERSONAL TOUCH

A virtual tour of your training facility can offer more than just visual stimulation. According to the Aberdeen Group, marketers who use video grow revenue 49% faster than non-video users. A well-crafted video tour can be a powerful marketing tool. It can demonstrate your school's commitment to advanced technology, the caliber of your instructors, and the rigor of your training—factors that can persuade a prospect to choose your flight school over competitors.

MARKETERS WHO USE VIDEO GROW REVENUE **49%** FASTER THAN NON-VIDEO USERS.

Incorporating video tours into your marketing strategy can be a game-changer for flight and ground schools. Not only do they provide visual stimulation, they can also significantly impact your bottom line. As stated earlier, marketers who use video grow revenue 49% faster than non-video users. Let's explore why video tours are so crucial and how you can leverage them effectively.

Building Trust and Credibility

One of the key benefits of video tours is their ability to build trust and credibility with potential students. When a prospect can virtually step inside your training facility and get a firsthand look at your classrooms, training materials, aircraft, flight simulators, and instructors, it conveys transparency and authenticity. This can go a long way in alleviating any doubts or skepticism potential students may have about the quality of your training.

Showcasing Your Technological Advancements

For aviation training institutions, staying at the forefront of technology is vital. A well-crafted video tour can highlight your commitment to advanced technology and innovation. You can showcase your state-of-the-art flight simulators, cutting-edge training materials, and any technological advancements that set your school apart from the competition. This positions your school as a leader in the field and attracts tech-savvy students.

Humanizing Your Brand

Beyond the physical aspects of your training facility, video tours offer an opportunity to humanize your brand. You can introduce potential students to your instructors, giving them a sense of the people who will guide them on their aviation journey. This personal touch can forge an emotional connection that resonates with prospects.

Planning Your Virtual Tour

To make the most of video tours, meticulous planning is essential. Start by creating a detailed storyboard that outlines the different aspects you want to showcase. Consider including segments on classrooms, training materials, aircraft, flight simulators, and, most importantly, the instructors. This comprehensive list will serve as your roadmap during the video production process.

> TO MAKE THE MOST OF VIDEO TOURS, **METICULOUS PLANNING IS ESSENTIAL.**

Crafting a Compelling Narrative

Once you have your list of elements to showcase, it's time to craft a compelling narrative for your virtual tour. Think about who will guide the tour – it could be an enthusiastic instructor or a satisfied student. Their role is crucial in conveying the story and creating a connection with the audience. A well-structured narrative can keep viewers engaged and informed throughout the tour.

Production Quality Matters

Investing in the production quality of your video tour is paramount. High-definition visuals and clear audio are essential to create a professional and engaging tour. Consider investing in professional equipment or hiring skilled videographers or video production teams to ensure the best possible quality.

> **A KEY BENEFIT OF VIDEO TOURS IS THEIR ABILITY TO BUILD TRUST AND CREDIBILITY WITH POTENTIAL STUDENTS.**

Promoting Your Video Tour

Once your video tour is ready, it's time to promote it effectively. Share it across your digital platforms, including your website, social media, and email marketing campaigns. If you're working with influencers, ask them to share it on their channels as well, to increase distribution beyond your current followers.

Tracking Engagement and ROI

TO GAUGE HOW EFFECTIVE YOUR VIDEO TOUR IS, **TRACK ENGAGEMENT METRICS.**

To gauge the effectiveness of your video tour, track engagement metrics such as views, watch time, and click-through rates. Analyze how the video tour impacts the decision-making process of prospective students. Do they spend more time on your website after viewing the tour? Are there increased inquiries or applications? By closely monitoring these metrics, you can measure the ROI of your video tour and make data-driven improvements.

Video tours are a powerful way for flight and ground schools to enhance their marketing efforts and drive ROI. They build trust, showcase technological advancements, humanize your brand, and offer a unique window into your training facility. By planning, crafting, and promoting your video tour effectively, you can leave a lasting impression on potential students and ultimately contribute to the growth and success of your institution.

BENEFITS OVER TRADITIONAL MEDIA

Unlike a brochure or a static webpage, a video can capture dynamic aspects of your training facility, like an instructor passionately explaining a concept on a smartboard, or a pilot-in-training navigating a flight simulator. Such real-world scenarios provide aspirants a glimpse of what they can expect, increasing the likelihood of them taking the next step in the enrollment process.

Estimating Your Video Tour Budget

Creating a captivating video tour for your flight or ground school requires careful budgeting to ensure you achieve the desired quality and impact. To help you plan effectively, we'll break down the budgeting process into three tiers: Good, Better, and Best. Each tier represents a different level of production quality and equipment, along with estimated costs (subject to change based on various factors like location and equipment upgrades).

Good

For a budget-conscious approach that still delivers a quality video tour, consider the following equipment and estimated costs:

1. Camera

A good-quality DSLR camera or a high-end smartphone with good video capabilities can suffice. **Estimated cost: $500 - $1,000**

2. Tripod

A basic tripod to stabilize your shots. **Estimated cost: $50 - $100**

3. Microphone

A lavalier microphone for clear audio recording.
Estimated cost: $50 - $100

4. Lighting

Natural light or basic LED lighting setups. **Estimated cost: $50 - $150**

5. Editing Software

Entry-level video editing software for post-production.
Estimated cost: $100 - $300

6. Miscellaneous

Additional expenses for props, transportation, and basic video effects.
Estimated cost: $100 - $200

Total Estimated Budget (Good): $850 - $1,850

A VIDEO CAN CAPTURE **THE DYNAMIC ASPECTS** OF YOUR TRAINING FACILITY.

Better

For a higher level of production quality and more professional results, consider investing in the following equipment and estimated costs:

1. Camera

A mid-range DSLR or mirrorless camera with 4K video capabilities.
Estimated cost: $1,000 - $2,500

2. Tripod

A sturdy tripod with fluid head for smoother camera movements. **Estimated cost: $150 - $300**

3. Microphone

A shotgun microphone for improved audio quality. **Estimated cost: $200 - $500**

4. Lighting

Professional-grade LED lighting kits with adjustable color temperature. **Estimated cost: $300 - $800**

5. Editing Software

A premium video editing software with advanced features. **Estimated cost: $300 - $600**

6. Miscellaneous

Additional budget for better-quality props, locations, and visual effects. **Estimated cost: $200 - $400**

Total Estimated Budget (Better): $2,150 - $5,100

TELLING A **COMPELLING STORY** THROUGH YOUR VIDEO IS CRUCIAL.

Best

For top-tier production quality that rivals industry standards, consider the following equipment and estimated costs:

1. Camera

A high-end cinema camera or DSLR with cinematic features. **Estimated cost: $5,000 - $10,000+**

2. Tripod

A professional video tripod with fluid head and advanced features. **Estimated cost: $1,000 - $2,500**

3.Microphone

A high-quality shotgun microphone with additional audio recording equipment. **Estimated cost: $1,000 - $3,000**

4. Lighting

Full professional lighting setups, including softboxes, diffusers, and color correction tools. **Estimated cost: $1,500 - $5,000**

5. Editing Software

Premium video editing software with advanced post-production capabilities. **Estimated cost: $500 - $1,200**

6. Miscellaneous

Significant budget for top-notch props, locations, visual effects, and professional actors or presenters. **Estimated cost: $1,000 - $5,000+**

Total Estimated Budget (Best): $9,000 - $26,700+

Remember that these estimated costs are subject to change based on your specific needs, location, and the availability of equipment. It's essential to allocate budget wisely based on your production goals and the quality you aim to achieve for your video tour. Additionally, consider hiring a professional videographer or production team for the "Better" or "Best" tiers to ensure the highest production value and a polished final product.

POORLY MADE VIDEOS CAN HAVE THE OPPOSITE EFFECT AND DETER POTENTIAL STUDENTS AND CAN **SEVERLY HARM YOUR SCHOOL'S REPUTATION.**

BEST PRACTICES

While the merits of video marketing are evident, the impact depends largely on the quality and strategic planning of your video content. Here are some best practices to ensure your virtual tours and other video content are both engaging and effective.

Quality is Key

First and foremost, invest in high-quality video production. Poorly made videos can have the opposite effect and deter potential students. Ensure good lighting, clear audio, and a stable camera. Given the sophistication involved in aviation training, a subpar video can severely harm your school's reputation.

Storytelling

Telling a compelling story through your video is crucial. Potential students should not only see your facilities but also understand the culture, the enthusiasm and expertise of the instructors, and the career possibilities that await them. A well-crafted story can help viewers see themselves as part of your institution, making them more likely to enroll.

SEO and Distribution

Your video is useless if it doesn't reach your target audience. Utilize SEO techniques, such as properly tagging your video and writing a descriptive transcript, to increase its visibility on platforms like YouTube. Also, ensure that the video is prominently featured on your website and shared across all social media platforms—by your brand and possibly by influencers—for maximum reach.

VIDEO MARKETING CHECKLIST

- ☐ **Plan Ahead:** Create a storyboard and script before filming.
- ☐ **Invest in Quality:** High-quality video and audio are non-negotiable.
- ☐ **Tell a Story:** Make sure your video has a compelling narrative.
- ☐ **SEO:** Use keywords, tags, and transcripts to make your video easily searchable.
- ☐ **Distribution:** Share the video across all platforms, including your website and social media, and through influencers.
- ☐ **Analyze:** Use video analytics to measure engagement and ROI.

VIDEO MARKETING IS A PERSONAL CONNECTION

By adhering to these best practices and utilizing the immense potential of video marketing, flight and ground schools can greatly improve their chances of converting prospective students into enrollments.

15

SHARE KNOWLEDGE ABOUT AVIATION TRAINING AND BUILD AN AUDIENCE WITH PODCASTING

CHAPTER FIFTEEN

Podcasting

BUILD AUTHORITY AND CREDIBILITY FOR YOUR SCHOOL

Podcasting is an increasingly popular medium that has witnessed a surge in both creators and consumers. Edison Research's 2022 Infinite Dial Report found, 41% of Americans above the age of 12 listen to podcasts regularly. In the realm of education and specialized training, podcasting can serve as a unique platform to engage current and prospective students. For flight and ground schools looking to enhance their digital marketing strategy, podcasting presents a virtually untapped channel for ROI and growth.

Podcasting can be particularly useful for complex subjects like aviation training. It offers an informal yet impactful way to disseminate knowledge, showcase expertise, and engage with a wider audience. While other schools are busy deploying conventional digital marketing strategies, a well-orchestrated podcast series could provide you with a significant competitive advantage.

PODCASTING

THE PERSONAL TOUCH

WHY PODCASTING FOR AVIATION TRAINING

You might wonder why a podcast is necessary for a niche like aviation training. The answer lies in the medium's ability to offer rich, engaging content that can be consumed passively. Imagine a student being able to listen to an episode about the nuances of flight safety while driving, or an aspiring pilot gaining valuable industry insights during a workout.

80%

OF PODCAST LISTENERS CONSUME ALL OR MOST OF EACH EPISODE THEY START.

According to a study by Podcast Insights, 80% of listeners consume all or most of each episode they start, a statistic that outpaces most video content. A podcast can feature interviews with industry experts, success stories from alumni, deep dives into aviation science, and Q&A sessions addressing prospective students' common concerns and questions.

Expert Interviews

A podcast can provide a platform to interview experienced pilots, air traffic controllers, and other aviation professionals. These interviews offer valuable information and establish your school's industry connections, thereby improving credibility and attracting potential enrollees.

Alumni Stories

What better way to showcase your training program's efficacy than to feature alumni success stories? According to Nielsen's Trust in Advertising report, 92% of consumers trust recommendations from people they know, and alumni narratives act as such personalized recommendations for your potential students.

Best Practices for Podcasting

Podcasting isn't just about recording an audio file and publishing it online; it's a multifaceted approach that requires careful planning and execution. Here are some best practices to get the most out of this medium.

STRATEGICALLY PLANNING YOUR PODCAST CONTENT

Creating a successful podcast for your flight school starts with a well-thought-out content strategy. To ensure your podcast effectively engages your target audience and aligns with your marketing goals, follow these steps:

A SUCCESSFUL PODCAST STARTS WITH **A WELL-THOUGHT-OUT CONTENT STRATEGY.**

1. Define Your Podcast's Objectives

Begin by clearly defining the primary objectives of your podcast. What purpose will it serve for your flight school? Is it designed to educate aspiring pilots, inspire aviation enthusiasts, or provide answers to common questions prospective students might have? Having a well-defined purpose will serve as the foundation for all your podcast content.

For instance, if your primary goal is to attract new student pilots, your podcast might focus on providing valuable insights into the training process, sharing success stories from your alumni, or discussing the latest developments in aviation training.

2. Identify Your Target Audience

Understanding your target audience is crucial for crafting content that resonates with them. Consider the demographics, interests, and pain points of your ideal listeners. Are you primarily targeting aspiring pilots, seasoned aviators, or aviation enthusiasts?

By creating a listener persona and gaining insights into their needs and preferences, you can tailor your podcast episodes to address their specific interests and challenges. This targeted approach increases the chances of attracting and retaining a loyal listener base.

3. Plan Your Podcast Episodes

Once you've established clear objectives and identified your target listener, it's time to plan your podcast episodes. Your content plan should include:

- **Episode Topics:** Brainstorm a list of relevant and engaging topics related to aviation training, industry insights, career opportunities, or any areas of interest to your target audience.

- **Episode Format:** Determine the format of your episodes. Will you have interviews with experienced pilots or instructors, solo discussions, or a combination of both? Variety can keep your content fresh and appealing.
- **Episode Length:** Decide on the ideal episode length. Over 50% of podcasts are between 20 and 60 minutes, per cloud-based podcast hosting solution Buzzsprout.[23] While some podcasts thrive with short, digestible episodes, others benefit from in-depth, longer-form content. Consider what works best for your audience and content.
- **Release Schedule:** Establish a consistent release schedule, whether it's weekly, biweekly, or monthly. Consistency helps build anticipation and keeps your audience engaged.
- **Episode Flow:** Plan the structure of each episode, including the introduction, main content, and conclusion. Having a clear outline ensures your podcast is organized and easy to follow.
- **Guests and Interviews:** If your podcast includes guest interviews, start identifying potential guests within the aviation industry or related fields. Reach out to them and schedule recording sessions in advance.

4. Craft Engaging Content

When creating your podcast episodes, prioritize engaging and valuable content. Provide insights, tips, stories, or expertise that your audience can't easily find elsewhere. Keep your content informative, entertaining, and relevant to your defined objectives and audience. Consider incorporating storytelling techniques, real-life examples, and practical advice to captivate your listeners and keep them coming back for more.

By following these strategic steps, you'll be well-prepared to plan and produce podcast content that effectively serves your flight school's goals, resonates with your target audience, and elevates your marketing efforts. Remember that podcasting is an ongoing journey, and continuously assessing and adjusting your content strategy based on listener feedback and performance metrics will lead to long-term success.

MONETIZATION

While the primary aim of your podcast may be to attract new students, it could also serve as an additional revenue stream. Consider sponsorship deals with aviation equipment suppliers, book publishers, or other entities interested in reaching your target audience.

DISTRIBUTION AND PROMOTION

Leverage existing channels, such as your school's social media pages, e-newsletters, and website, to promote the podcast. It would also be beneficial to list your podcast on various platforms to reach a wider audience. According to Podcast Hosting, distributing your podcast on multiple platforms can increase listenership by up to 30%.

TECHNICAL ASPECTS

Quality matters in podcasting as much as it does in any other medium. In fact, poor audio quality is one of the top reasons listeners abandon a podcast, according to a report by MusicOomph. Use professional-grade microphones, soundproof your recording environment, and invest in good quality editing software.

DISTRIBUTING YOUR PODCAST ON VARIOUS PLATFORMS CAN INCREASE LISTENERSHIP BY UP TO **30%**

Estimating Your Podcasting Equipment Budget

Creating a professional podcast requires the right equipment. Here, we'll provide budget estimates for three tiers: Good, Better, and Best. Each tier represents different levels of podcasting equipment quality and estimated costs (subject to change based on brand choices and specific needs).

Good

For those just starting with podcasting or on a tight budget, you can achieve decent audio quality with the following equipment:

1. Microphone

A USB condenser microphone offers good audio quality for beginners. **Estimated cost: $50 - $100**

2. Headphones

A pair of closed-back headphones for monitoring audio while recording. **Estimated cost: $30 - $70**

3. Pop Filter

A basic pop filter to reduce plosive sounds. **Estimated cost: $10 - $20**

4. Recording Software

Free or low-cost recording and editing software. **Estimated cost: $0 - $50**

Total Estimated Budget (Good): $90 - $240

Better

For improved audio quality and more professional results, consider the following equipment:

YOUR PODCAST MAY BE INTENDED TO ATTRACT NEW STUDENTS, BUT **IT COULD ALSO SERVE AS** AN ADDITIONAL REVENUE STREAM.

1. Microphone

A dynamic or condenser XLR microphone for better sound. **Estimated cost: $100 - $300**

2. Audio Interface

A quality audio interface to connect XLR microphones to your computer. **Estimated cost: $100 - $300**

3. Headphones

Professional closed-back studio headphones for accurate monitoring. **Estimated cost: $100 - $200**

4. Boom Arm/Stand

A boom arm or microphone stand for stability and flexibility. **Estimated cost: $30 - $100**

5. Pop Filter

A higher-quality pop filter for cleaner audio. **Estimated cost: $20 - $50**

6. Acoustic Treatment

Basic foam panels or isolation shields for soundproofing. **Estimated cost: $50 - $150**

7. Editing Software

Premium podcasting software with advanced features. **Estimated cost: $100 - $300**

Total Estimated Budget (Better): $500 - $1,100

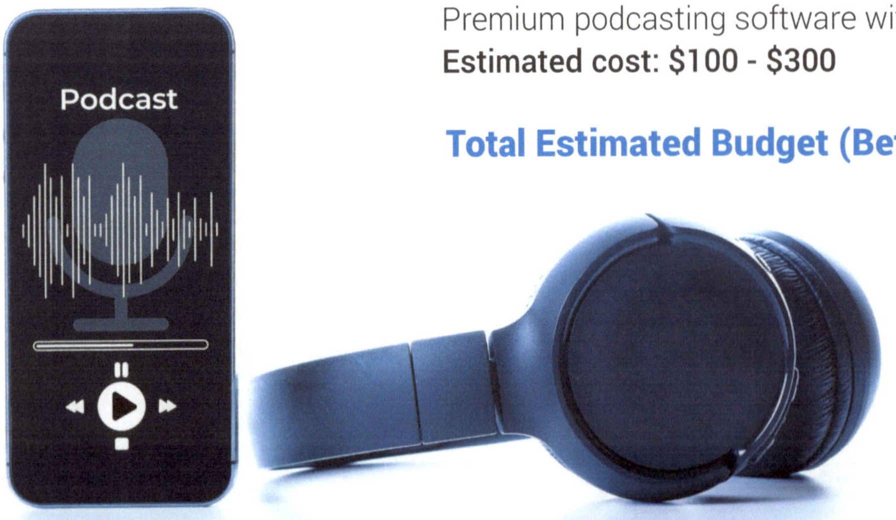

Best

For top-tier podcasting quality and professional results, consider investing in the following equipment:

1. Microphone

A high-end XLR condenser microphone from a reputable brand. **Estimated cost: $300 - $1,000+**

2. Audio Interface

A premium audio interface with advanced features and sound quality. **Estimated cost: $300 - $700**

3. Headphones

Professional-grade, closed-back headphones with exceptional sound fidelity. **Estimated cost: $200 - $500**

4. Boom Arm/Stand

A premium, adjustable boom arm or microphone stand. **Estimated cost: $100 - $300**

5. Pop Filter

A top-quality pop filter for the cleanest audio possible. **Estimated cost: $50 - $100**

6. Acoustic Treatment

Comprehensive acoustic treatment with bass traps, diffusers, and acoustic panels. **Estimated cost: $300 - $1,000+**

7. Editing Software

Industry-standard podcasting software for advanced post-production. **Estimated cost: $200 - $500**

8. External Mixer

A high-quality mixer for precise audio control. **Estimated cost: $300 - $1,000+**

Total Estimated Budget (Best): $1,650 - $4,300+

PODCASTING IS AN ONGOING JOURNEY. CONTINUE TO ASSESS AND ADJUST YOUR CONTENT STRATEGY BASED ON LISTENER FEEDBACK AND PERFORMANCE METRICS **FOR LONG-TERM SUCCESS.**

Your specific budget will depend on your podcasting goals and quality standards. While the "Good" tier is budget-friendly and suitable for beginners, investing in the "Better" or "Best" tiers will deliver superior audio quality and a more professional podcasting experience. Keep in mind that these estimated costs are subject to change based on your equipment preferences and any additional accessories you may need.

PODCASTING CHECKLIST

- ☐ **Content Planning:** Clearly define your objectives and plan your episodes accordingly.
- ☐ **Technical Quality:** Invest in top-notch recording and editing equipment.
- ☐ **Monetization:** Look for sponsorship opportunities to add a revenue stream.
- ☐ **Distribution:** Use existing channels for promotion and consider multiple platforms for hosting.
- ☐ **Metrics:** Measure success through listener count, engagement rate, and ROI.

VIDEO MARKETING IS A PERSONAL CONNECTION

When done right, podcasting can act as an essential marketing tool for flight and ground schools, providing a platform for education and engagement, as well as an additional revenue generation and growth.

16

MOBILE APPS ARE AN INVALUABLE TOOL TO REACH AN AUDIENCE WHO IS ALWAYS ON THE GO

CHAPTER SIXTEEN

Mobile App Marketing

ON-THE-GO TRAINING INFORMATION

In an increasingly mobile-centric world, having a dedicated application for your flight or ground school can provide unparalleled advantages. According to App Annie, users spent 3.5 trillion hours on mobile apps in 2021, showcasing the growing reliance on mobile platforms for both entertainment and education. The advantage of mobile apps in education is their ability to provide targeted information, user-specific learning paths, and real-time updates—all accessible on-the-go. For aviation training schools aiming for impressive ROI and growth, mobile app marketing can offer numerous untapped opportunities.

MOBILE APP MARKETING

THE IMPORTANCE OF A MOBILE APP

YOUR PODCAST MAY BE INTENDED TO ATTRACT NEW STUDENTS, BUT **IT COULD ALSO SERVE AS** AN ADDITIONAL REVENUE STREAM.

Why Mobile Apps Matter

Mobile applications offer a direct pipeline to your prospective and current students, serving as a hub for educational content, school updates, faculty interaction, and even virtual tours of your training facilities. The presence of a dedicated mobile app can also elevate your school's brand, providing a professional, tech-savvy image that attracts the younger generation.

Feature-Rich Apps

A feature-rich app with functionalities such as interactive training modules, real-time flight tracking for students, and appointment scheduling with instructors, can be an invaluable asset for your flight school. According to a study by the Mobile Marketing Association, feature-rich educational apps have a 30% higher retention rate compared to basic informational apps.

User Engagement

Engaging your user base is crucial for the success of any mobile application. Push notifications, for example, can serve as gentle reminders for upcoming classes or new course material, driving user engagement and fostering a committed learning environment.

MARKETING STRATEGIES FOR MOBILE APPS

App Store Optimization (ASO)

Just like SEO for websites, App Store Optimization is crucial for the visibility of your app. A report by Sensor Tower found, 70% of app downloads come from organic searches within app stores. Well-researched keywords, compelling visuals, and positive reviews can significantly boost your app's visibility.

User Acquisition and Retention

Attracting new users and retaining existing ones are equally critical. Promotional offers for app downloads, exclusive in-app content, and loyalty programs can all drive user acquisition and retention. A report by CleverTap indicates that a well-executed user retention strategy can increase customer lifetime value by up to 25%.

Monetization Strategies

While the primary objective of your mobile app may be educational, that doesn't mean you can't monetize it effectively. In-app purchases offering advanced courses, simulator experiences, or specialized training materials can provide an additional revenue stream, contributing to your overall ROI.

MEASURING SUCCESS

Key Metrics

Measuring the success of your mobile app involves tracking various KPIs including download rates, user engagement, session times, and conversion rates. Monitoring these metrics allows you to fine-tune your marketing strategies and app functionalities.

ROI Analysis

Determining the ROI of your mobile app involves a careful study of the revenue generated against the investment made in development and marketing. According to a survey by Soomla, educational apps that kept a close eye on ROI metrics were 60% more successful in achieving their financial goals.

A/B Testing

As with other digital marketing strategies, A/B testing can be an invaluable tool in optimizing your mobile app's performance. Different layouts, features, and notification styles can be tested to determine what maximizes user engagement and, ultimately, ROI.

PODCASTING IS AN ONGOING JOURNEY. CONTINUE TO ASSESS AND ADJUST YOUR CONTENT STRATEGY BASED ON LISTENER FEEDBACK AND PERFORMANCE METRICS **FOR LONG-TERM SUCCESS.**

MOBILE APP MARKETING CHECKLIST

- ☐ **Feature Planning:** Decide on which functionalities would be most beneficial for your users.
- ☐ **App Store Optimization:** Invest in ASO to enhance your app's visibility.
- ☐ **User Acquisition:** Plan effective strategies for both user acquisition and retention.
- ☐ **Monetization:** Consider various ways to monetize the **app effectively.**
- ☐ **Analytics:** Use robust metrics to measure success and adjust strategies accordingly.

BUILD TOOLS WITH MOBILE APP MARKETING

With comprehensive planning, precise execution, and continuous performance monitoring, mobile app marketing can serve as a significant pillar for your flight school's digital marketing strategy. Its potential for enhancing both the educational experience and your institution's ROI is immense and growing with each passing year.

PART IV

MEASURE & MONITOR

17

CAREFUL AND REGULAR DATA ANALYSIS IS CRITICAL FOR YOUR MARKETING SUCCESS

CHAPTER SEVENTEEN

Data Analytics

WHAT GETS MEASURED GETS IMPROVED

The significance of data analytics in today's business landscape is akin to the importance of a compass for a pilot navigating through the skies. Analytics empower aviation training schools to make informed decisions, optimize marketing campaigns, and enhance customer experience, thereby driving ROI and growth. According to Forbes, companies who adopt data-driven marketing are six times more likely to be profitable year-over-year. This chapter aims to explore the various facets of data analytics specific to flight and ground schools, reviewing the kind of metrics that should be monitored and how to translate these numbers into actionable insights.

DATA ANALYTICS

UNDERSTANDING METRICS

What to Measure

COMPANIES THAT SET DATA-DEFINED GOALS ACHIEVE A **19%** INCREASE IN PROFITABILITY

Deciding what metrics to track can often be overwhelming, given the plethora of data points available. KPIs like website visits, engagement rates, and customer lifetime value should be on your radar. These are indicative of your marketing effectiveness and can help shape future strategies.

Metric-driven Goals

Setting metric-driven goals can guide your team's focus and align everyone on quantifiable objectives. These goals could range from increasing website conversion rates to reducing customer acquisition costs. A study by MIT suggests companies that set data-defined goals achieve a 19% increase in profitability.

Real-time Monitoring

The real-time analytics provided by tools like Google Analytics and Adobe Analytics allow for timely adjustments in your marketing campaigns. They help in spotting trends and anomalies that could be opportunities or threats, respectively.

ANALYTICS TOOLS AND PLATFORMS

Free vs. Paid Tools

While there are free analytics tools that offer basic functionalities, investing in a paid platform can significantly amplify the insights you gain. Tools like Tableau and Salesforce offer robust features like data visualization and predictive analytics. Gartner states that, companies investing in advanced analytics tools see an average ROI increase of 27%.

Data Integration

Most modern analytics platforms offer seamless integration with other data sources like CRM systems, email marketing platforms, and even social media analytics. This makes for a centralized data hub that can give a 360-degree view of your marketing efforts.

Security and Compliance

Given the sensitivity of data, ensuring its security is paramount. Most premium analytics tools are GDPR compliant and have multiple layers of security protocols. Protect your data and your reputation.

TRANSLATING DATA INTO ACTION

The Art of Segmentation

One of the most effective ways to utilize analytics is through segmentation. Data can help segment your customer base into different categories based on behavior, preferences, or stage in the customer journey. This leads to more personalized and effective marketing campaigns. A study by Campaign Monitor found that segmented campaigns drive a 760% increase in revenue.

Experimentation and Optimization

The ability to A/B test different strategies enables data-driven optimization. Whether it's the subject lines of your email marketing campaigns or the layout of your landing pages, A/B testing helps you understand what resonates with your audience. Businesses that use A/B testing have a 33% chance of beating their growth goals, according to VWO.

Automation and Reporting

Data analytics, when combined with automation tools, can result in an ultra-efficient marketing machine. Scheduled reports, automated alerts, and predictive analytics can save time and improve ROI. A report by McKinsey indicates that businesses automating data analytics see a 15% increase in marketing efficiency.

CONCLUSION

The advent of data analytics has transformed how flight and ground schools approach marketing. By understanding what to measure, how to measure it, and most importantly, how to act upon these measurements, aviation training institutions can significantly increase their ROI and spur growth. In an industry as competitive as this, failing to capitalize on the power of analytics would be akin to ignoring your instruments while flying through a storm.

18

ENSURE THAT YOUR VISITORS TURN INTO STUDENTS USING CRO

CHAPTER EIGHTEEN

Conversion Rate Optimization

TURNING VISITORS INTO STUDENTS

Conversion Rate Optimization (CRO) is the art and science of converting website visitors into engaged students, whether that means signing up for a newsletter, enrolling in a course, or simply taking an online tour of your training facilities. Just as a pilot relies on accurate instruments to achieve a smooth landing, your digital marketing strategy must incorporate well-placed tactics to guide potential students through the enrollment funnel effectively. According to Adobe, companies with the strongest omni-channel customer engagement strategies enjoy a 10% year over year growth, a 10% increase in average order value, and a 25% increase in close rates. This chapter will walk you through the pivotal aspects of CRO that can significantly improve your ROI and growth.

CONVERSION RATE OPTIMIZATION

THE FUNDAMENTALS OF CRO

The Importance of Landing Pages

Your landing page is often the first point of interaction with a potential student. Its design, content, and user interface play critical roles in capturing interest. Landing pages must be tailored to specific campaigns and audiences for maximum effectiveness. WordStream asserts that companies with 40 or more landing pages get 12 times more leads than those with 5 or fewer.

61%

OF USERS ARE UNLIKELY TO RETURN TO A MOBILE SITE IF THEY HAD TROUBLE ACCESSING IT.

CTAs: The Final Push

A compelling call-to-action can make or break your conversion rates. Placement, language, and even color can influence the effectiveness of a CTA. Studies from HubSpot reveal that personalized CTAs perform 202% better than generic ones.

MOBILE OPTIMIZATION AND USER EXPERIENCE (UX)

The Mobile Surge

In an increasingly mobile-centric world, overlooking mobile optimization is a costly mistake. Google states that 61% of users are unlikely to return to a mobile site if they had trouble accessing it, with 40% visiting a competitor's site instead.

User Experience: The Unseen Engine

An intuitive, user-friendly experience not only enhances your site's usability but also significantly boosts conversion rates. Forrester reports that a well-designed user interface could raise your conversion rates up to 200%, and a better UX design could yield conversion rates up to 400%.

OVERLOOKING MOBILE OPTIMIZATION IS A COSTLY MISTAKE.

BEHAVIORAL ANALYSIS AND PERSONALIZATION

Monitoring User Behavior

Understanding how users interact with your website offers invaluable insights into improving CRO. Heatmaps, session replays, and click tracking provide detailed behavioral analysis that can highlight both strong points and bottlenecks on your website.

Personalization: The Secret Sauce

Incorporating personalized content or offers based on user behavior or demographics can lift your conversion rates exponentially. According to Econsultancy, 93% of companies see an uplift in conversion rates from personalization.

A/B TESTING AND CONTINUOUS IMPROVEMENT

A/B Testing: The Reality Check

A/B testing allows you to experiment with different variables, from landing page designs to email subject lines, to see which options yield the highest conversions. Optimizely found that companies that are continuously A/B testing generate 16% more revenue compared to those that don't.

CONVERSION RATE OPTIMIZATION **ISN'T A ONE-TIME EFFORT** BUT AN ONGOING PROCESS.

The Loop of Continuous Improvement

The field of digital marketing is ever-evolving, and what worked yesterday may not be as effective today. Hence, your fight school's mantra should always be to test, analyze, and optimize. A study by Bain & Company shows that companies who excel at continuous improvement grow revenues 2.2 times faster than their industry peers.

CONCLUSION

Conversion rate optimization isn't a one-time effort but an ongoing process. Every element, from your landing pages to mobile optimization and personalization, needs to be continually scrutinized and optimized for effectiveness. By applying CRO best practices, aviation training centers stand to increase enrollment rates and boost the ROI and growth of their marketing campaigns significantly.

19

UNDERSTAND THE VALUE OF CUSTOMER RETENTION TO YOUR GROWTH AND PROFIT

CHAPTER NINETEEN

Customer Retention and Lifetime Value

BUILD LONG-TERM PROFIT BY FOCUSING ON RETENTION

While acquisition is crucial, retention is what determines your aviation training center's long-term success. Statistics from Invesp indicate that it costs five times as much to acquire a new customer than to keep an existing one. Moreover, increasing customer retention rates by just 5% can grow profits by 25% to 95%. This chapter focuses on methods and strategies to improve customer retention and explores the concept of customer lifetime value (CLV), both crucial metrics for ensuring robust ROI and sustained growth in the aviation training sector.

LIFETIME VALUE

UNDERSTANDING CUSTOMER RETENTION

THE COST OF ACQUIRING A NEW STUDENT **IS FAR GREATER** THAN RETAINING AN EXISTING ONE.

Why Retention Matters

The cost of acquiring a new student is far greater than retaining an existing one. But the advantages go beyond cost-effectiveness; a retained customer is more likely to refer new students, thus amplifying your marketing efforts. A study by Yotpo found that, returning customers spend more over time, boosting your CLV metrics and ultimately, your ROI.

Strategies for Retention

Email campaigns, loyalty programs, and excellent customer service are just some of the tools at your disposal for improving retention. Exclusive offers or "alumni-only" training sessions can work wonders in keeping your students engaged. An American Express survey found that 68% of customers churn because of poor service, emphasizing the value of a well-trained support team.

UNLOCKING THE POWER OF CUSTOMER LIFETIME VALUE (CLV)

Calculating CLV

Understanding the CLV of your students can dramatically impact your marketing strategies and allocation of resources. By knowing what a customer is worth over the course of their relationship with your business, you can optimize marketing spend and target the most profitable customer segments.

Customer Lifetime Value

$$\text{CLV} = \left[\text{Customer Value} \times \text{Average Customer Lifespan} \right]$$

Leveraging CLV for ROI

According to HBR, increasing customer retention rates will inevitably result in a spike in customer lifetime value. By focusing on long-term relationships, you can enhance your ROI through upselling, cross-selling, and customer referrals.

TECHNOLOGICAL TOOLS FOR RETENTION AND CLV

CRM Systems

Customer Relationship Management (CRM) systems are invaluable in tracking customer behavior, preferences, and engagement levels. $8.71 is earned for every dollar spent on CRM, per Nucleus Research, making it a wise investment for boosting retention and CLV.

Data Analytics

Sophisticated analytics tools can reveal patterns and trends in customer behavior, allowing you to make data-driven decisions. Software like Google Analytics can track key metrics like repeat visits, average time on site, and specific interactions that lead to conversion and retention.

CONCLUSION

Customer retention and CLV are not merely add-ons to your marketing strategy; they are integral components that sustain and grow your aviation training center. By investing in quality service, leveraging technology, and making data-driven decisions, you can cultivate lasting relationships with your students. Not only does this improve your bottom line, but it also builds a community around your brand, offering compounding returns on your initial investment.

20

HOW TO UNDERSTAND THE PERFORMANCE OF YOUR DIGITAL MARKETING STRATEGIES.

CHAPTER TWENTY

ROI AND KPIs

FINANCIAL METRICS FOR FLIGHT SCHOOLS

Understanding the performance of your digital marketing strategies is essential to your success. It's crucial for flight schools to track specific metrics to assess the Return on Investment (ROI) and set appropriate Key Performance Indicators (KPIs). Studies by Econsultancy indicate that companies allocating more than 50% of their marketing budget to measurement and analytics are three times more likely to outperform their competitors in sales. This chapter will elucidate the essential financial metrics flight and ground schools should monitor to measure ROI effectively and set useful KPIs.

KEY PERFORMANCE INDICATORS (KPIS)

Identifying Relevant KPIs

Setting relevant KPIs starts with a thorough understanding of your objectives. Are you looking to increase student enrollment, improve retention, or maybe boost the visibility of a specific course? A study by the Content Marketing Institute revealed that the most effective marketers are those who document their strategy, including KPIs.

Implementing Tracking Mechanisms

ONCE YOU'VE IDENTIFIED THE KPIs, TRACKING THEM IS **THE NEXT LOGICAL STEP.**

Once you've identified the KPIs, tracking them is the next logical step. Utilizing advanced analytics tools can offer a wealth of insights. Google Analytics, for instance, provides granular data on website performance, user engagement, and conversion rates.

ROI METRICS

Financial Indicators

Revenue per student, cost per acquisition, and profit margin are some of the critical financial indicators to track for ROI. Companies that focus on calculating ROI are 1.6 times more likely to receive higher budgets for the following fiscal year, according to the Journal of Interactive Marketing.

Aligning Marketing Strategies with ROI

Incorporating ROI metrics into your broader marketing strategy enables you to allocate resources more efficiently. By focusing your efforts on high-performing channels and campaigns, you can improve your overall profitability, as noted in a Forrester Research report.

DATA-DRIVEN DECISION-MAKING

Role of Analytics

Your KPIs and ROI metrics are only as good as the data they are based on. Employing analytics tools can facilitate in-depth insights into customer behavior and campaign effectiveness. According to a report by McKinsey companies that invest in analytics have a 23% higher likelihood of outperforming their competitors in terms of customer satisfaction and a 19% higher chance of exceeding profitability margins.

Making Iterative Improvements

Digital marketing is continually evolving. Therefore, frequent evaluation and adjustment of your KPIs and ROI metrics are essential. A dynamic approach ensures that you are not wasting resources on outdated methods and are always in sync with current market trends.

CONCLUSION

The significance of ROI and KPIs in the digital marketing landscape for flight and ground schools is immense. Focused efforts on tracking, measuring, and interpreting these financial metrics can offer a competitive edge. Keeping an eye on ROI and KPIs allows for agile strategies that adapt to market conditions and ensure maximum profitability.

PART V

RESOURCES

21

COMPARE THE BENEFITS OF CREATING YOUR OWN TEAM OR CHOOSING AN AGENCY

CHAPTER TWENTY ONE

In-House vs. Agency

THE PROS AND CONS OF BUILDING A TEAM IN-HOUSE OR HIRING AN AGENCY

The dilemma of whether to build an in-house marketing team or to hire an agency is often a pivotal decision that can shape the trajectory of your flight school's growth. According to Gartner, 29% of the marketing budget is typically allocated to in-house labor costs, whereas outsourcing consumes about 22%. Both approaches have their merits and demerits, and the right choice can significantly affect your ROI and growth. This chapter reviews the intricacies of both options, providing insights to help you make an informed decision for your aviation training business.

IN-HOUSE VS. AGENCY

THE PROS AND CONS OF AN IN-HOUSE TEAM

Pros of In-House

An in-house marketing team offers deep industry and organizational understanding and alignment with your company culture. Studies by MarketingProfs indicate that in-house teams often have a better grasp of brand identity and long-term objectives.

Cons of In-House

CHOOSING BETWEEN AN IN-HOUSE TEAM AND AN AGENCY **IS A MULTI-FACETED DECISION.**

However, the drawback of an in-house team is the higher fixed costs in salaries, benefits, and training. Additionally, the learning curve for specialized skills like SEO or PPC can be steep and time-consuming.

THE PROS AND CONS OF HIRING AN AGENCY

Pros of Agency

Agencies bring specialized expertise and a diversified skill set to the table. A report by Deloitte suggests agencies often deliver quicker ROI due to their focused expertise in varied marketing channels.

Cons of Agency

The downside can be the higher cost per project and the risk of misalignment with your brand's voice and objectives. Agencies may juggle multiple clients, potentially diluting the focus on your specific needs.

FACTORS TO CONSIDER: BUDGET, EXPERTISE, AND TIME

Budget Considerations

If your marketing budget is tight, the variable costs of an agency might be more appealing than the fixed costs of a full-time team (Gartner, 2020).

Expertise Required

For highly specialized tasks, the advanced skills and tools that an agency offers can provide a quicker path to ROI.

Time Sensitivity

In-house teams may require time for ramp-up and training, whereas agencies can often hit the ground running, enabling quicker campaign launches.

CASE STUDIES: SUCCESS AND FAILURES

In-House Success

A study from CMO Council showed that one flight school saw a 35% increase in student enrollments within six months of building an in-house team, citing better alignment with company goals as a significant factor.

Agency Success

Conversely, according to eMarketer, another aviation training center experienced a 40% ROI within the first quarter of hiring an agency, attributing the success to the agency's specialized skill set.

Failures to Learn From

Both approaches have had their share of failures, often due to lack of proper planning, unrealistic expectations, or miscommunication.

CONCLUSION

Choosing between an in-house team and an agency is a multifaceted decision involving budget, expertise, time, and alignment with organizational goals. Both have proven successful in different contexts, as indicated by real-world case studies. The pivotal factor often lies in understanding your specific needs, constraints, and long-term vision. According to a comprehensive survey by the Content Marketing Institute, businesses that align their marketing choices with their strategic goals are 30% more likely to experience higher ROI. Therefore, a thorough evaluation is crucial for making the optimal choice for your flight or ground school.

UNDERSTAND YOUR SPECIFIC NEEDS, CONSTRAINTS, AND LONG-TERM VISION.

22

UTILIZE ALL OF THE TOOLS AVAILABLE TO YOU TO DRIVE YOUR FLIGHT SCHOOL MARKETING

CHAPTER TWENTY TWO

Essential Tools and Software

YOUR TOOLBOX FOR DIGITAL MARKETING SUCCESS

The right tools and software can act as a force multiplier for your flight or ground school's marketing initiatives. According to a survey by MarTech Today, businesses that effectively leverage marketing technology saw an average ROI improvement of 50%. From customer relationship management (CRM) platforms to data analytics software, this chapter will outline the essential tools that can streamline operations, enhance efficiency, and boost ROI.

ESSENTIAL TOOLS AND SOFTWARE

MARKETING TOOLBOX MUST-HAVES

CRM SOFTWARE: YOUR CUSTOMER MANAGEMENT HUB

Why CRM is Crucial

CRM software acts as a hub for all interactions with your student base and prospects. Salesforce states, companies that implement CRM solutions witness sales increases of up to 29%.

COMPANIES THAT IMPLEMENT CRM SOLUTIONS WITNESS SALES INCREASES OF UP TO **29%**

Recommended CRM Software

Tools and software are constantly changing. As soon as we go to print, things will change. We have created a resource for you where we go more in-depth at pilottrainingreviews.com/book. Tools like Salesforce, HubSpot, and Zoho offer specialized features tailored for educational institutions, which can be particularly useful for flight and ground schools. Register for free at pilottrainingreviews.com/book to receive your free resource list.

SEO TOOLS: OPTIMIZE FOR VISIBILITY

The ROI of SEO

Effective SEO can bring in a constant stream of organic traffic. A study by BrightEdge shows that organic search accounts for 53% of all website traffic, a source that shouldn't be ignored.

SEO Tools to Consider

Platforms like SEMrush, Ahrefs, and Moz offer in-depth analytics and recommendations tailored to enhance your website's visibility.

EMAIL MARKETING PLATFORMS: COMMUNICATION THAT CONVERTS

Why Email Marketing?

According to a report by Campaign Monitor, for every $1 spent on email marketing, the average ROI is $42, making it one of the most effective channels for marketing.

FOR EVERY **$1 SPENT** ON EMAIL MARKETING, **THE AVERAGE ROI IS $42**

Software Choices

Mailchimp, Constant Contact, and SendinBlue are among the best platforms offering a range of features from A/B testing to analytics, imperative for optimizing your email campaigns.

SOCIAL MEDIA MANAGEMENT TOOLS: YOUR DIGITAL LOUDSPEAKER

ROI from Social Media

According to Sprout Social, 74% of consumers rely on social networks to guide purchasing decisions. Effective social media management tools can help you tap into this potential.

Tools to Use

Hootsuite, Buffer, and Sprout Social offer features like content scheduling, analytics, and audience segmentation that can significantly impact your school's social reach.

ANALYTICS AND DATA VISUALIZATION: MAKING SENSE OF NUMBERS

The Importance of Analytics

Data analytics can provide invaluable insights into customer behavior and campaign effectiveness. According to IBM, businesses that use analytics are five times more likely to make faster decisions.

Tools for Analytics

Google Analytics, Tableau, and Microsoft Power BI are top-of-the-line analytics tools that can help your flight school understand data and subsequently improve ROI.

CONCLUSION

The software and tools you select will act as the backbone of your marketing operations, affecting every aspect of your campaigns, from planning to execution to analysis. The Forrester Wave report suggests that integrating these tools seamlessly can lead to an average efficiency increase of 15-20%.

CHOOSE YOUR ARSENAL OF DIGITAL TOOLS

As a flight or ground school aiming for superior ROI and growth, equipping yourself with the right arsenal of digital tools is not just advisable—it's indispensable.

GLOSSARY

COMMON TERMS & DEFINITIONS

Artificial Intelligence (AI)

Artificial intelligence (AI) is a technology that helps computers make decisions and solve problems similarly to how humans do.

Augmented Reality (AR)

Augmented Reality (AR) is a technology that overlays digital information, like images or animations, onto the real world through devices such as smartphones or AR glasses.

Attribution Modeling

Attribution modeling is a method used in digital marketing to determine how credit for sales or conversions is assigned to different touchpoints in a customer's journey.

B2B Marketing

Business to Business Marketing

Customer Acquisition Cost (CAC)

Total expense incurred to acquire a new customer, including marketing expenses, advertising, wages, and other related costs.

Customer Lifetime Value (CLV)

Customer Value (sum of all benefits to the company) multiplied by the Average Customer Lifespan

Cost Per Acquisition (CPA)

Amount of money spent to acquire a new customer through a specific marketing campaign or channel. Specifically, it indicates how much they spend for each new student enrollment resulting from a particular advertising or promotional effort.

Customer Relationship Management (CRM)

A system or software used to manage a company's interactions with current and potential customers, organizing information about them and tracking their history with the company.

Conversion Rate Optimization (CRO)

The process of increasing the percentage of website visitors who take a desired action, such as signing up for a course or making a purchase.

Calls to Action (CTA)

A prompt or instruction on a website or in marketing material that encourages the audience to take a specific action, such as "Sign Up," "Learn More," or "Book Now."

Data Analytics

Data analytics involves examining raw data to draw conclusions and uncover patterns or insights, often using statistical or computational methods.

Earned Media Value

The value attributed to publicity gained through promotional efforts other than paid advertising.

Email Automation

The process of sending out emails to subscribers or customers automatically, based on triggers or schedules.

Engagement Rate

A metric that measures the level of interaction content receives from an audience, such as likes, shares, and comments.

Google Analytics (GA)

A free tool by Google that tracks and reports website traffic and user behavior.

Inbound Marketing

A marketing strategy that attracts customers by creating valuable content tailored to them.

Key Performance Metrics

Specific measurements that show how effectively a company is achieving its business objectives.

Keyword Density

The percentage of times a keyword appears on a web page compared to the total number of words on that page.

Keyword Strategy

A plan for choosing and implementing keywords to drive traffic to a website.

Key Performance Indicator (KPI)

A measurable value that indicates how well a company is achieving its key objectives.

Lead Capture Mechanisms

Tools or methods used to collect information from potential customers, like sign-up forms.

Long-tail Keywords

Specific and longer keyword phrases that visitors are more likely to use when they're closer to making a purchase.

Lifetime Customer Value

The total revenue a business can expect from a single customer throughout their relationship.

Meta Tags

Snippets of text in a page's code that describe its content, often used by search engines.

Omni-channel

An approach to sales and marketing that provides customers with a seamless experience across all channels.

Outbound Marketing

Traditional marketing methods that push products or services to a broad audience, like TV ads or direct mail.

Retargeting

A digital marketing strategy that targets ads to people who have previously visited a website or interacted with a brand.

Schema Markup

A type of microdata that helps search engines understand the content on web pages and provide better search results.

Search Engine Optimization (SEO)

The practice of optimizing a website to increase its visibility in search engine results.

Search Engine Results Pages (SERPs)

The pages displayed by search engines in response to a query.

UTM parameters

Tags added to URLs to track the performance of campaigns and content in analytics tools.

User Experience (UX)

The overall experience a user has with a product, system, or service.

Virtual Reality (VR)

A computer-generated simulation of a three-dimensional environment that can be interacted with in a seemingly real way using special equipment.

PART VI

PLANNING WORKBOOK

INCLUDED IN THIS WORKBOOK SECTION:

1. ACTION STEPS
2. MARKETING STRATEGY CHECKLISTS
3. MARKETING STRATEGY TEMPLATE
4. BUDGET WORKSHEET
5. SOCIAL MEDIA CHART
6. SOCIAL MEDIA PLANNING CALENDAR
7. SAMPLE MARKETING CALENDAR

MARKETING STRATEGY PLANNING

ACTION STEPS

1. Conduct surveys asking your current students or interested parties about their goals, challenges, and demographics.
2. Analyze website and social media analytics to determine the characteristics of your most engaged audience.
3. Implement AI chatbots on your website to provide personalized course recommendations based on individual queries.
4. Use segmented email lists to send out targeted promotions or information.
5. Determine which social media platforms are most most used by your target audience.
6. Create a social strategy that provides a balance of educational and promotional content.
7. Develop a posting schedule that aligns with when your target audience is most active.
8. Create blog posts and video content that directly address the challenges and questions your target audience has about aviation training.
9. Share testimonials or case studies that reflect similar journeys to those of your personas.
10. Regularly review and update your personas.
11. Keep an eye on customer reviews and social media discussions to understand emerging trends or shifts in perception.

MARKETING STRATEGY CHECKLISTS

AI MARKETING CHECKLIST

- ☐ Investigate AI solutions suitable for customer interaction and data analysis.
- ☐ Explore the feasibility of incorporating VR and AR into your marketing strategies.
- ☐ Assess the applicability of blockchain technology in your business operations.
- ☐ Revisit your SEO strategy to include voice search optimization. Artificial Intelligence.

HIGH-CONVERTING WEBSITE CHECKLIST

- ☐ Clear, Actionable CTAs: Make sure your CTAs stand out and tell the visitor exactly what action they should take next.
- ☐ Personalized Landing Pages: Create different landing pages for different audience segments.
- ☐ Trust Indicators: Showcase reviews, testimonials, and certifications.
- ☐ Speed Optimization: Test your website's speed and improve it for both desktop and mobile devices.
- ☐ Mobile Responsiveness: Ensure your website is easily navigable on mobile devices.
- ☐ Lead Capture Mechanisms: Integrate sign-up forms, quizzes, or other lead capture tools into your site.
- ☐ Ongoing Analytics and Testing: Regularly review analytics data and run A/B tests to optimize performance.

ON-PAGE OPTIMIZATION CHECKLIST FOR IMPROVED SEO

1. Keyword Research

- ☐ Conduct thorough keyword research to identify relevant and high-impact keywords for your aviation training content.
- ☐ Consider long-tail keywords, which are specific and can attract more targeted traffic.

2. Keyword Placement

- ☐ Utilize keywords strategically in the following elements:

- **Meta Titles:** Craft unique and keyword-rich meta titles for each page, making them compelling and under 60 characters.
- **Meta Descriptions:** Create engaging meta descriptions that include keywords and encourage click-throughs (under 155-160 characters).
- **Header Tags (H1, H2, H3, etc.):** Structure your content with clear headers, incorporating keywords in the H1 and H2 tags as naturally as possible.

3. Original and High-Quality Content

- ☐ Ensure that your content is unique, informative, and well-researched.
- ☐ Avoid duplicate content, as it can negatively impact your SEO efforts.
- ☐ Regularly update and refresh content to keep it relevant and valuable.

4. Keyword Density

- ☐ Maintain a reasonable keyword density (typically 1-2%) to avoid over-optimization, which can harm your rankings.
- ☐ Use keywords naturally within the content, ensuring readability and user engagement.

5. URL Structure

- ☐ Craft user-friendly, descriptive URLs that include keywords when applicable.
- ☐ Use hyphens (-) to separate words in URLs for better readability.

6. Internal Linking

- ☐ Implement internal links to connect related content within your website.
- ☐ Use descriptive anchor text that includes relevant keywords.

7. Optimize Images

- ☐ Compress and optimize images to improve page load times.
- ☐ Use descriptive alt text for images, including keywords where appropriate.

8. Mobile Optimization

- ☐ Ensure your website is mobile-responsive for a seamless user experience on all devices.
- ☐ Test your site's mobile-friendliness using tools like Google's Mobile-Friendly Test.

9. Page Speed

- ☐ Optimize your website's load speed by minimizing unnecessary scripts and using browser caching.
- ☐ Check your site's speed with tools like Google PageSpeed Insights and GTmetrix.

10. User Experience (UX)

- ☐ Prioritize user-friendly navigation and a clean, intuitive design.
- ☐ Improve user experience to reduce bounce rates and encourage longer on-page time.

11. Schema Markup

- ☐ Implement schema markup to provide search engines with structured data about your content, making it more appealing in search results.

12. Regular Updates

- ☐ Continuously monitor and update your on-page optimization efforts based on performance data and changing search trends.

13. Monitoring and Analysis

- ☐ Use tools like Google Analytics and Google Search Console to track the performance of your optimized pages.
- ☐ Adjust your strategy based on the insights gained from these tools.

14. Meta Description Optimization

- ☐ Craft compelling meta descriptions that not only include keywords but also entice users to click through.
- ☐ Follow best practices for meta description length and format.

15. User Intent

- ☐ Consider user intent when optimizing content. Ensure that your content matches what users are looking for when they search with specific keywords.

16. External Linking

- ☐ Include relevant external links to authoritative sources when appropriate, enhancing the credibility of your content.

17. HTTPS

- ☐ Secure your website with HTTPS to protect user data and improve search engine rankings.

18. Social Sharing

- ☐ Implement social sharing buttons on your content to encourage sharing and increase visibility.

OFF-PAGE OPTIMIZATION CHECKLIST FOR IMPROVED SEO

1. Backlink Strategy

- ☐ Develop a comprehensive backlink strategy aimed at acquiring high-quality and relevant backlinks from authoritative websites within the aviation industry.

2. Link Outreach

- ☐ Identify reputable aviation blogs, flight training forums, industry news websites, and other related platforms that could potentially link to your content.

3. Quality over Quantity

- ☐ Prioritize the quality of backlinks over their quantity. A few high-authority backlinks are more valuable than numerous low-quality ones.

4. Content Creation

- ☐ Produce high-quality and valuable content that naturally attracts backlinks from other websites. Content can include blog posts, guides, infographics, and industry reports.

5. Guest Blogging

- ☐ Consider guest blogging opportunities on aviation-related websites. Ensure that your guest posts are informative, well-researched, and provide value to the audience.

6. Link Building Outreach

- ☐ Reach out to webmasters, bloggers, and site owners to propose collaboration opportunities, such as guest posting, link exchanges, or content partnerships.

7. Broken Link Building

- ☐ Identify broken links on aviation-related websites and offer to replace them with relevant, high-quality content from your site.

8. Social Media Promotion

- ☐ Promote your content on social media platforms to increase its visibility and the likelihood of it being shared by others in the aviation community.

9. Online Forums and Communities

- ☐ Actively participate in aviation forums, communities, and discussion boards. Provide valuable insights, answers, and solutions while discreetly linking back to your site when relevant.

10. Monitor Mentions

- ☐ Use tools like Google Alerts or Mention to monitor online mentions of your brand or content. Engage with those who mention your site and potentially request backlinks when appropriate.

11. Directory Listings

- ☐ Ensure that your aviation training business is listed in relevant online directories and industry-specific listings.

12. Press Releases

- ☐ Publish press releases about noteworthy events or achievements related to your aviation training business. Distribute them through reputable channels to earn backlinks from news websites.

13. Collaborate with Influencers

- ☐ Partner with aviation influencers or experts in the field for co-authored content, interviews, or collaborations. Their endorsement can attract valuable backlinks.

14. Monitor Competitor Backlinks

- ☐ Analyze the backlink profiles of your competitors using tools like Ahrefs or SEMrush. Identify opportunities to acquire similar or better-quality backlinks.

15. Link Diversity

- ☐ Aim for a diverse range of backlinks from various sources, including blogs, news sites, forums, and educational institutions.

16. Avoid Black Hat Tactics

- ☐ Strictly avoid black hat SEO practices such as buying backlinks, link farms, or spammy link-building schemes. These can result in penalties from search engines.

17. Analyze Backlink Quality

- ☐ Regularly assess the quality and relevance of your existing backlinks. Disavow or request removal of low-quality or spammy backlinks that may harm your site's reputation.

18. Track Backlink Performance

- ☐ Monitor the impact of your backlink-building efforts on your site's rankings and organic traffic. Use tools like Google Search Console and SEO analytics platforms to track progress.

BUSINESS PROFILE CHECKLIST

Google My Business (GMB) Optimization

- ☐ **Create or Claim Your Listing:** Start by creating a Google My Business account or claiming an existing one. Ensure all information is accurate and complete.
- ☐ **Accurate NAP:** Consistency is key; make sure your Name, Address, and Phone number (NAP) are consistent across all online platforms.
- ☐ **Categories and Attributes:** Select relevant categories and attributes that accurately describe your aviation training school.
- ☐ **Photos and Videos:** Regularly update your GMB profile with high-quality images and videos showcasing your facilities, instructors, and aircraft.

- ☐ **Customer Reviews:** Encourage satisfied students to leave reviews on your GMB listing. Respond to reviews, both positive and negative, to engage with your audience.
- ☐ **Posts and Updates:** Utilize the GMB Posts feature to share news, events, and updates about your school, helping to keep your listing active and engaging.

Bing Places for Business

- ☐ **Claim Your Bing Listing:** Create or claim your listing on Bing Places for Business, mirroring the information on your GMB profile.
- ☐ **Complete Your Profile:** Fill in all available fields, including a detailed description of your aviation training services.
- ☐ **Images and Videos:** Upload compelling visuals that highlight your school's unique offerings.
- ☐ **Customer Reviews:** Encourage students to leave reviews on your Bing listing, as positive reviews can boost your search visibility.

Apple Maps Connect

- ☐ **Listing Verification:** Sign into Apple Maps Connect and verify your listing through the verification code sent to your business phone number.
- ☐ **Accurate Information:** Ensure that all details, such as your school's name, address, and contact information, are accurate and up to date.
- ☐ **Categories:** Select appropriate categories that align with your aviation training services.
- ☐ **Imagery:** Upload eye-catching images that showcase your school's facilities, aircraft, and training environment.

LOCAL SEO CHECKLIST FOR AVIATION TRAINING SCHOOLS

1. Claim and Verify Listings

- ☐ Create or claim your listings on Google My Business, Bing Places for Business, and Apple Maps Connect.

2. Consistent NAP

- ☐ Maintain a consistent Name, Address, and Phone number (NAP) across all listings and directories.

3. Complete Profiles

- ☐ Fill out all available fields with accurate information, including a detailed description of your aviation training services.

4. High-Quality Visuals

- ☐ Regularly update your profiles with high-resolution images and videos that reflect the quality of your school.

5. Customer Reviews

- ☐ Encourage satisfied students to leave reviews on all platforms, and respond to reviews promptly.

6. Categories and Attributes

- ☐ Select relevant categories and attributes that accurately represent your aviation training school.

7. Regular Updates

- ☐ Keep your profiles active by posting news, events, and updates about your school.

8.Local Keywords

- ☐ Optimize your website and content for local keywords related to aviation training and your location.

9. Mobile Optimization

- ☐ Ensure your website is mobile-friendly, as many local searches occur on mobile devices.

10. Local Link Building

- ☐ Build local backlinks from aviation-related businesses and directories.

11. Monitoring and Analytics

- ☐ Regularly monitor your local SEO performance using tools like Google Analytics and Search Console.

12. Stay Informed

- ☐ Stay up to date with changes in local SEO algorithms and best practices to maintain and improve your rankings.

CHECKLIST FOR IMPLEMENTING A SEO STRATEGY

- ☐ Keyword Research: Use tools like Google's Keyword Planner or SEMrush to identify relevant keywords.
- ☐ On-Page Optimization: Modify meta titles, meta descriptions, and header tags to include target keywords.
- ☐ Off-Page Optimization: Earn backlinks from authoritative domains in the aviation and training sectors.
- ☐ Content Marketing: Consistently produce high-quality, original content that is relevant to your target audience.
- ☐ Local SEO: Optimize your Google My Business account and focus on local keywords.
- ☐ Technical SEO: Ensure website speed, mobile optimization, and secure, accessible websites.
- ☐ Analytics and Monitoring: Use tools like Google Analytics and Google Search Console for consistent monitoring and optimization.

BLOGGING FOR AVIATION TRAINING BUSINESSES CHECKLIST

- ☐ **Quality Over Quantity:** Prioritize high-quality, informative content over frequent posting. Strive for consistency without sacrificing content quality.
- ☐ **Posting Frequency:** Choose a posting frequency that aligns with your resources and audience expectations, whether it's weekly, bi-weekly, or monthly.

- ☐ **Keyword Research:** Conduct thorough keyword research to identify trending topics and phrases relevant to aviation training. Incorporate these keywords naturally into your content.
- ☐ **Diverse Content Types:** Embrace a variety of content formats, such as videos, infographics, podcasts, and case studies, to cater to different audience preferences.
- ☐ **Audience-Centric Approach:** Craft content that addresses the specific needs, concerns, and interests of your target audience, including topics like FAA regulations, pilot training processes, and career paths in aviation.
- ☐ **Promotion and Distribution:** Maximizing Reach
- ☐ **Email Marketing:** Notify your email subscribers of new blog posts through regular newsletters or updates that summarize recent content.
- ☐ **Social Media:** Choose the most relevant social media platforms for your audience and share blog posts. Engage with your audience and foster discussions around your content.
- ☐ **Guest Posting:** Collaborate with aviation-related websites, forums, or influencers to guest post and expand your reach.
- ☐ **Content Repurposing:** Transform blog content into different formats, such as videos or infographics, to appeal to diverse audience preferences.
- ☐ **Engage and Interact:** Encourage comments and discussions on your blog posts. Respond promptly to foster a sense of community around your content.
- ☐ **Measuring Success:** Analytics and Insights
- ☐ **Google Analytics:** Use Google Analytics to track website traffic, user behavior, and conversion rates resulting from your blog posts.
- ☐ **Conversion Tracking:** Set up conversion tracking within Google Analytics to measure specific actions taken by visitors after reading your blog posts.
- ☐ **Keyword Performance:** Monitor how your blog posts rank for target keywords using tools like Moz and SEMrush.
- ☐ **Audience Feedback:** Actively seek and analyze feedback from your audience through surveys or social media polls to understand their preferences.

MARKETING STRATEGY CHECKLISTS

CONTENT MARKETING CHECKLIST

- ☐ Blogs: Regularly publish value-driven SEO-optimized article.
- ☐ Videos: Use various platforms to distribute different types and formats of video content.
- ☐ Podcasts: Establish a regular podcast to engage your audience.
- ☐ Infographics: Utilize graphics to simplify complex topics.
- ☐ Webinars: Host regular webinars.
- ☐ Social Media Posts: Maintain an active presence across relevant social channels.
- ☐ E-books and Whitepapers: Offer downloadable, value-driven long-form content.
- ☐ E-newsletters: Use email marketing to keep your audience updated and informed.

EMAIL MARKETING CHECKLIST

- ☐ Segmentation: Divide your email list based on key attributes such as the stage of training or areas of interest.
- ☐ Content Planning: Develop a content calendar to keep your email marketing consistent and valuable.
- ☐ Analytics: Track key metrics like open rates, click-through rates, and conversion rates to gauge the success of your campaigns.
- ☐ A/B Testing: Regularly run A/B tests to optimize various elements of your emails.
- ☐ Automation: Implement automated email series for specific actions like welcoming new students or post-training follow-ups.
- ☐ Legal Compliance: Ensure you adhere to all legal requirements related to email marketing.

SOCIAL MEDIA MARKETING CHECKLIST

- ☐ Select appropriate platforms based on your target audience.
- ☐ Develop a documented content strategy.
- ☐ Track KPIs regularly for actionable insights.

- ☐ Utilize both organic and paid social media strategies.
- ☐ Adhere to legal guidelines specific to social media marketing.

PAY-PER-CLICK ADVERTISING CHECKLIST

- ☐ Choose the most appropriate PPC platform based on your target audience.
- ☐ Conduct in-depth keyword research, considering the nuances of user intent.
- ☐ Craft compelling ad copy and optimize landing pages for conversions.
- ☐ Decide on a bidding strategy aligned with your campaign objectives.
- ☐ Monitor key metrics regularly for actionable insights.
- ☐ Adhere to legal guidelines specific to PPC advertising.

AFFILIATE MARKETING CHECKLIST

- ☐ Identify potential affiliates that align with your brand and target audience.
- ☐ Determine a commission structure that benefits both parties.
- ☐ Implement precise tracking methods for transparent attribution.
- ☐ Utilize affiliate marketing for dual benefits—direct revenue and SEO.
- ☐ Always adhere to legal guidelines and ensure affiliates do the same.

INFLUENCER MARKETING CHECKLIST

- ☐ Explore advanced influencer marketing strategies such as long-term collaborations and co-creation of content.
- ☐ Learn from real-world case studies to glean insights into the effectiveness of influencer marketing.
- ☐ Stay ahead of emerging trends, including the rise of nano-influencers and sustainability influencers.
- ☐ Measure the ROI of influencer marketing campaigns using KPIs like engagement rate, CTR, conversion rate, and CPA.

- ☐ Ensure legal compliance and transparency in influencer partnerships by adhering to FTC guidelines and crafting comprehensive contracts.

VIDEO MARKETING CHECKLIST

- ☐ Plan Ahead: Create a storyboard and script before filming.
- ☐ Invest in Quality: High-quality video and audio are non-negotiable.
- ☐ Tell a Story: Make sure your video has a compelling narrative.
- ☐ SEO: Use keywords, tags, and transcripts to make your video easily searchable.
- ☐ Distribution: Share the video across all platforms, including your website and social media, and through influencers.
- ☐ Analyze: Use video analytics to measure engagement and ROI.

PODCASTING CHECKLIST

- ☐ Content Planning: Clearly define your objectives and plan your episodes accordingly.
- ☐ Technical Quality: Invest in top-notch recording and editing equipment.
- ☐ Monetization: Look for sponsorship opportunities to add a revenue stream.
- ☐ Distribution: Use existing channels for promotion and consider multiple platforms for hosting.
- ☐ Metrics: Measure success through listener count, engagement rate, and ROI.

MOBILE APP MARKETING CHECKLIST

- ☐ Feature Planning: Decide on which functionalities would be most beneficial for your users.
- ☐ App Store Optimization: Invest in ASO to enhance your app's visibility.
- ☐ User Acquisition: Plan effective strategies for both user acquisition and retention.
- ☐ Monetization: Consider various ways to monetize the app effectively.
- ☐ Analytics: Use robust metrics to measure success and adjust strategies accordingly.

MARKETING STRATEGY TEMPLATE

EXECUTIVE SUMMARY

Objective: Briefly outline the primary goals of the marketing strategy.

Target Audience: Define the target customer segments.

Budget: Provide an overview of the budget allocation.

SITUATION ANALYSIS

Market Overview

Market Trends: Describe current market trends relevant to the business.

Competitive Analysis: List key competitors and their strengths and weaknesses.

MARKETING STRATEGY TEMPLATES

SWOT Analysis

Strengths:

____

____

Weaknesses:

____

____

Opportunities:

____

____

Threats:

____

____

Marketing Objectives

Objective 1:

____

____

Objective 2:

____

____

Objective 3:

____

____

Target Audience

Segment 1:

- Characteristics:

  ___

  ___

- Needs:

  ___

  ___

Segment 2:

- Characteristics:

  ___

  ___

- Needs:

  ___

  ___

Marketing Channels

1. Website and SEO:

- Action Items:

  ___

  ___

2. Social Media:

- Action Items:

  ___

  ___

MARKETING STRATEGY TEMPLATES

3. **Email Marketing:**
    - Action Items:

      ____

      ____

Budget and Resources

Channel Allocation:

____

____

Resource Requirements:

____

____

KPIs and Metrics

KPI 1:

____

____

KPI 2:

____

____

Implementation Timeline

Phase 1:

____

____

Monitoring and Evaluation

Review Dates:

Evaluation Criteria:

CONCLUSION

Summarize the key points and next steps.

Note: *This is a template and should be customized according to specific business needs.*

Categories

- **Marketing Channels:** Budget allocation for Website and SEO, Social Media, Email Marketing, PPC Advertising, and Affiliate Marketing.
- **Content Creation:** Budget for Blog Posts, Video Production, Infographics, and Webinars.
- **Public Relations:** Budget for Press Releases, Events and Sponsorships, and Influencer Partnerships.
- **Analytics and Tools:** Budget for SEO Tools, Email Marketing Tools, and Social Media Management Tools.
- **Miscellaneous:** A contingency budget for unexpected expenses.

Recommendations for Monetization and Improvement:

- **Dynamic Allocation:** Keep your budget flexible. As you measure ROI, reallocate funds to the most profitable channels.
- **Tool Integration:** Use integrated analytics tools to track spending and ROI in real-time. This will help you make data-driven decisions.
- **Sponsored Content:** Given your focus on aviation training, consider allocating budget for sponsored content on relevant platforms. This can drive high-quality traffic and also serve as an additional revenue stream.
- **Retargeting Budget:** Allocate a portion of your PPC and Social Media budget for retargeting campaigns. These generally offer a higher ROI.
- **Influencer Partnerships:** Given the community-driven nature of PilotTrainingReviews.com, influencer partnerships can be a cost-effective way to gain credibility and reach.
- **Premium Features:** For PilotTrainingReviews.com, consider a budget for developing premium, value-added features that can be monetized.
- **Cost Control:** Always keep a contingency budget for unforeseen opportunities or challenges. This ensures you can adapt without compromising other activities.

MARKETING STRATEGY PLANNING

BUDGET WORKSHEET

Category	Sub-Category	Allocated Budget	Actual Spend	ROI
Marketing Channels	Website and SEO	$	$	$
	Social Media	$	$	$
	Email Marketing	$	$	$
	PPC Advertising	$	$	$
	Affiliate Marketing	$	$	$
Content Creation	Blog Posts	$	$	$
	Video Production	$	$	$
	Infographics	$	$	$
	Webinars	$	$	$
Public Relations	Press Releases	$	$	$
	Evens and Sponsorships	$	$	$
Analytics	SEO Tools	$	$	$
	Email Marketing Tools	$	$	$
	Social Media Management Tools	$	$	$
	Contingency - Miscellaneous	$	$	$
Total		**$**	**$**	**$**

	YOUTUBE	INSTAGRAM	TIKTOK	(FORMERLY TWITTER)	FACEBOOK	LINKEDIN
ROLE	Educate & Enhance SEO	Entertain & Engage	Entertain & Inspire	Update & Educate	Entertain & Drive Leads	Credentialize
KEY DEMOGRAPHIC	15 - 35 51% M 49% F	18 - 24 52% M 48% F	18 - 24 54% M 46% F	18 - 29 66% M 34% F	25 - 34 56% M 44% F	30 - 39 57% M 43% F
AVG. DAILY USAGE	48.7 min.	33.1 min.	53.8 min.	34.1 min.	30.9 min.	7.27 min.
TYPES OF CONTENT	Course Overviews & Testimonials	Inspirational Videos & Images	Inspirational Videos	Topical Comments & Updates	Updates & Links	Thought Leadership & Company News
CONTENT LENGTH	2 min.	Stories - :15 Reels - :15 - : 60 Posts - 125 chars	7-15 sec.	71-100 chars	40-80 chars	100-140 chars
POSTING FREQUENCY	1x/week	1-2x/day	1-4x/day	3-4x/day	4-5x/day	1x/day

SOCIAL MEDIA DETAILS

The following channels are all viable ways for flight and ground schools to reach and engage with potential students:

- **YouTube** has the second highest MAUs at just over 2.5 billion1, and is popular among a wide age range, 15-35. Properly optimizing your videos helps potential students find them on YouTube and through Google searches.
- **Instagram** is evolving from photo-sharing to video-sharing. Reels, in particular, are becoming more widespread as a means to increase reach and engagement since they're discoverable by all users. Stories and posts are only seen by your followers.
- **TikTok** describes itself as an entertainment platform rather than a social platform. Users are there to discover trends and learn, rather than to connect, which is why "how-to" videos and tutorials perform well on TikTok.
- **X (formerly Twitter)** offers your brand a way to engage in real-time conversations and to find and index information on niche topics like "instrument ratings" by using hashtags. It can be a valuable source of website traffic and leads.
- Despite some recent declines, **Facebook** is still the dominant social media platform worldwide, with more than 2.9 billion monthly active users (MAUs). (Statista)
- As the leading professional and B2B networking platform, **LinkedIn** is highly effective for lead generation. Its users span a variety of age ranges, enabling aviation schools to reach targets for different levels of training, from beginners to those seeking advanced certifications. (Sprout Social)

These are some well-known social platforms we aren't recommending:

- **Snapchat's** user base has expanded beyond teens since the introduction of Snapchat+, a paid subscription that offers "a collection of exclusive, experimental, and pre-release features" for about $3.99/month.1 However, its recent AR experiments have been unsuccessful, and it continues struggling to secure ad revenues, so its longevity may be in question.
- Meta's new Twitter competitor, **Threads**, doesn't allow hashtag usage. This makes finding content focused on a specific topic challenging and requires a higher content volume to gain visibility.2 Daily active users spiked to 44 million just two days after its July 5th launch, but dropped to eight million DAUs (down 82%) by the end of July.
- **Reddit** users don't respond as well to ads or promoted content as users of other social media platforms. Its forum-style structure can be valuable for market research through social listening, especially for smaller businesses.

SOCIAL MEDIA PLANNING CALENDAR

Information to include:

- Type of Content
- Description
- Time of Posting

	MONDAY	TUESDAY	WEDNESDAY
LINKEDIN			
INSTAGRAM			
FACEBOOK			
TWITTER / X			
YOUTUBE			

THURSDAY	FRIDAY	SATURDAY	SUNDAY

SAMPLE MARKETING CALENDAR

Elements

- **Month and Week:** Specifies the time frame for each activity.
- **Marketing Activity:** Describes the specific task or campaign to be executed.
- **Channel:** Identifies the marketing channel to be used, such as Website and SEO, Content Marketing, Social Media, etc.
- **Target Audience:** Defines the customer segment you aim to target.
- **Goals:** Outlines the objectives of each marketing activity.
- **KPIs:** Lists the key performance indicators to measure the success of each activity.
- **Responsible Team Member:** Assigns the team member responsible for executing the activity.

Recommendations for Monetization and Improvement:

- **Seasonal Campaigns:** Consider adding seasonal or event-based campaigns to capitalize on high-traffic periods. For instance, a special campaign around aviation events could be lucrative.
- **A/B Testing:** Allocate some weeks for A/B testing of email campaigns, landing pages, or social media ads. This will help you optimize for better ROI.
- **User-Generated Content:** Plan weeks where you focus on encouraging and sharing user-generated content. This could be especially effective for PilotTrainingReviews.com.
- **Retargeting:** Include retargeting campaigns in your calendar to re-engage potential customers who have interacted with your platforms but haven't converted.
- **Content Syndication:** For high-performing content, allocate time for syndication across different platforms. This can extend the reach and lifespan of your content.
- **Analytics Review:** Regularly schedule weeks for a comprehensive review of analytics. This will help you understand what's working and what needs adjustment.
- **Monetization Strategies:** For PilotTrainingReviews.com, consider planning weeks dedicated to implementing and testing new monetization strategies, such as premium listings or value-added content.

MARKETING STRATEGY PLANNING

SAMPLE MARKETING CALENDAR

Month	Week	Marketing Activity	Channel	Target Audiences	Goals	KPIs	Responsibile Team Members
Jan	1	Website Audit	Website & SEO	All	Identify areas for improvement	Website Traffic	SEO Team
Jan	2	New Blog Post Release	Content Marketing	Student Pilots	Increase Engagment	Page Views	Content Team
Jan	3	Social Media Campaign	Social Media	Aviation Companies	Brand Awareness	Engagement Rate	Social Media Team
Jan	4	Email Newsletter	Email Marketing	Subscribers	Drive Traffic	Click-Through Rate	Social Media Team
Feb	1	PPC Campaign	PPC Advertising	Prospective Clients	Lead Generation	Conversion Rate	Advertising Team
Feb	2	Influencer Campaign	Public Relations	General Public	Increase Reach	Impressions	PR Team
Feb	3	SEO Update	Website and SEO	All	Improve Rankings	Search Engine Ranking	SEO Team
Feb	4	Webinar	Content Marketing	Student Pilots	Education & Engagement	Attendance Rate	Content Team

ENDNOTES

1 Pew Research Center. (n.d.). Internet/Broadband Fact Sheet. Retrieved September 23, 2023, from https://www.pewresearch.org/internet/fact-sheet/internet-broadband/

2 U.S. Census Bureau. (n.d.). Estimated Quarterly U.S. Retail Sales: Total and E-commerce. Retrieved September 23, 2023, from https://www.census.gov/retail/ecommerce.html

3 The Bazaar Voice. (July 7, 2021). Review criteria: What matters to shoppers most? Retrieved September 23, 2023, from https://www.bazaarvoice.com/blog/review-criteria-what-matters-to-shoppers-most/

4 Salesforce. (August 14, 2018). Shopper-First Retailing: New Research from 6,000 Consumers and 1.4 Billion Ecommerce Visits Reveals What Shoppers Actually Want. Retrieved September 23, 2023, from https://www.salesforce.com/blog/digital-shopper-first-retail-report-research-blog/

5 Accenture. (2018). Personalization Pulse Check | Accenture, "MAKING IT PERSONAL". Retrieved September 23, 2023, from https://www.accenture.com/_acnmedia/pdf-77/accenture-pulse-survey.pdf

6 BrandCamp Digital. (January 30, 2022). Understanding ROAS/ROI in Google Ads campaigns. Retrieved September 23, 2023, from https://www.brandcampdigital.com/post/understanding-roas-roi-in-google-ads-campaigns

7 Hubspot. (June 30, 2022). Customer Experience ROI: How to Convince Leadership It's Worth It. Retrieved September 23, 2023, from https://blog.hubspot.com/service/customer-experience-roi

8 Harvard Business Review. (November 2015). Customer Experience | The New Science of Customer Emotions. https://hbr.org/2015/11/the-new-science-of-customer-emotions

9 Forbes Advisor. (May 18, 2023). Top Social Media Statistics And Trends Of 2023. https://www.forbes.com/advisor/business/social-media-statistics/

10 McKinsey & Company. (January 2014). Using Customer Analytics to Boost Corporate Performance: Key insights from McKinsey's DataMatics 2013 survey. Retrieved September 23, 2023 from https://www.mckinsey.com/~/media/McKinsey/Business%20Functions/Marketing%20and%20Sales/Our%20Insights/Five%20facts%20How%20customer%20analytics%20boosts%20corporate%20performance/Datamatics.pdf

11 IDC. (August 2022). How Data Maturity and Product Analytics Improve Digital Experiences and Business Outcomes. Retrieved September 23, 2023, from heap.io/lp/ppc-how-data-maturity-drives-business-success

12 Optinmonster. (August 19, 2023). 18 Affiliate Marketing Statistics 2023 All Marketers Must Know. Retrieved October 20, 2023, from https://optinmonster.com/affiliate-marketing-statistics/#:~:text=Affiliate%20marketing%20has%20become%20a,industry%20could%20reach%20$40%20billion

13 Insider Intelligence. (October 9, 2023). 5 Charts Showing the Creator Economoy's Recent Evolution. Retrieved November 3, 2023, from https://www.insiderintelligence.com/content/5-charts-showing-creator-economy-s-recent-evolution

14 Statista. (August 29, 2023). Number of Social Media Users Worldwide from 2017 to 2027. Retrieved November 3, 2023, from https://www.statista.com/statistics/278414/number-of-worldwide-social-network-users/

15 Forbes. (May 17, 2018). Businesses Lose $75 Billion Due to Poor Customer Service. Retrieved November 3, 2023, from https://www.forbes.com/sites/shephyken/2018/05/17/businesses-lose-75-billion-due-to-poor-customer-service/?sh=26f72fae16f9

16 Demand Sage. (August 14, 2023). 67 Voice Search Statistics For 2023 (Fresh Data). Retrieved October 30, 2023 from https://www.demandsage.com/voice-search-statistics/

17 GoldenNewswire. (July 20, 2023). Voice Search Market to Record a CAGR of 19.2%, Speech Recognition Segment to be Largest Revenue - Generating Function Type Segment, 2023-2032. Retrieved November 3, 2023, from https://www.globenewswire.com/en/news-release/2023/7/20/2707992/0/en/Voice-Search-Market-to-Record-a-CAGR-of-19-2-Speech-Recognition-Segment-to-be-Largest-Revenue-Generating-Function-Type-Segment-2023-2032.html

18 Big Commerce. How to Use Customer Testimonials to Generate 62% More Revenue From Every Customer, Every Visit". Retrieved 30 October 2023 from https://www.bigcommerce.com/blog/customer-testimonials/

19 ahrefs blog. (November 25, 2011). Keywords. What They Are and How to Get Search Traffic From Them. Retrieved October 30, 2023 from https://ahrefs.com/blog/long-tail-keywords/

20 Optinmonster. (August 19, 2023). 18 Affiliate Marketing Statistics 2023 All Marketers Must Know. Retrieved October 30, 2023 from https://optinmonster.com/affiliate-marketing-statistics

21 Influencer Marketing Hub. (October 30, 2023). The State of Influencer Marketing 2023: Benchmark Report. Retrieved October 3, 2023 from https://influencermarketinghub.com/influencer-marketing-benchmark-report/

22 Social Pilot. (May 15, 2023). 100+ Surprising Video Marketing Statistics You Should Know in 2023. Retrieved October 3, 2023 from https://www.socialpilot.co/blog/video-marketing-statistics

23 Buzzsprout. (March 7, 2023). Podcast Statistics and Data. Retrieved October 3, 2023 from https://www.buzzsprout.com/blog/podcast-statistics.

www.ingramcontent.com/pod-product-compliance
Lightning Source LLC
LaVergne TN
LVHW070934210926
848482LV00003B/240
* 9 7 9 8 9 8 9 4 3 1 3 1 1 *